Fundamentals of Galaxy Dynamics, Formation and Evolution

T0136905

Fundamentals of Galaxy Dynamics, Formation and Evolution

Ignacio Ferreras

First published in 2019 by
UCL Press
University College London
Gower Street
London WC1E 6BT
Available to download free: www.ucl.ac.uk/ucl-press

ISBN: 978-1-911307-63-1 (Hbk.)
ISBN: 978-1-911307-62-4 (Pbk.)
ISBN: 978-1-911307-61-7 (PDF)
ISBN: 978-1-911307-64-8 (epub)
ISBN: 978-1-911307-65-5 (mobi)
ISBN: 978-1-911307-66-2 (html)
DOI: https://doi.org/10.14324/111.9781911307617

Contents

List of figures   ix
List of tables   xi
Preface   xiii
Acknowledgements   xv

1 An introduction to galaxy formation   1
   1.1 The main ingredients of a galaxy   1
   1.2 Observables   2
   1.3 Physical processes   13
   1.4 Stellar clusters   22
   1.5 A technical note on astronomical observations   22

2 The classical theory of gravitation   27
   2.1 Gravitational force   27
   2.2 The Kepler problem   29
   2.3 Potential theory   32
   2.4 Gravitational potential energy   35
   2.5 Potential/density pairs: A few fundamental cases   36
   2.6 Two-dimensional projection   41

3 A statistical treatment of stellar systems   43
   3.1 Phase space   43
   3.2 The distribution function   44
   3.3 Relaxation time   45
   3.4 Local and distant encounters   48
   3.5 Collisionless Boltzmann equation   51
   3.6 Isolating integrals: Jeans theorem   53

3.7 Examples of distribution functions 55
3.8 Jeans equations 60
3.9 The virial theorem 64
3.10 Beyond the collisionless Boltzmann equation: The Fokker-Planck equation 65

4 Understanding our Galaxy 70
4.1 General description of the Galaxy 70
4.2 Differential rotation in the Galaxy 74
4.3 Vertical motion 83
4.4 The collisionless Boltzmann equation in galactic coordinates 85
4.5 Application of Jeans equations 87
4.6 The potential of the Galaxy 89

5 Specific aspects of disc and elliptical galaxies 92
5.1 'Hot' versus 'Cold' dynamical systems 92
5.2 Scaling relations 94
5.3 Rotation versus 'pressure' in early-type galaxies 99
5.4 A brief introduction to spiral arms in disc galaxies 100

6 Galactic chemical enrichment 107
6.1 Nucleosynthesis and the formation of galaxies 107
6.2 General aspects of galactic chemical enrichment 108
6.3 Basic equations of galactic chemical enrichment 112
6.4 Chemistry as a cosmic clock 118

7 The growth of density fluctuations 121
7.1 A cosmology primer 122
7.2 Linear regime 129
7.3 Spherical collapse 135
7.4 Press-Schechter formalism 139
7.5 Correlation function 141
7.6 Cooling and the masses of galaxies 146

8 Smaller stellar systems: Stellar clusters 149
8.1 Open and globular clusters 149
8.2 Internal evolutionary effects 151
8.3 External effects: Tidal disruption 155
8.4 Cluster evaporation: King models 159

9 Larger stellar systems: Galaxy clusters 163
9.1 The most massive structures 163
9.2 X-ray measurements of the cluster mass 164

9.3 Gravitational lensing   166
9.4 Clusters and cosmology   170
9.5 Environment-related processes   171

Further reading   175
Index   177

List of figures

1.1	Panchromatic view of galaxy M81	3
1.2	Galaxy spectra from SDSS	4
1.3	Hubble's tuning fork diagram	7
1.4	Galaxy formation *in silico*	11
1.5	Colour-mass diagram	17
1.6	Abundance matching	18
2.1	Kepler's orbits	31
2.2	Newton's theorem	34
2.3	Projection of a mass distribution	41
3.1	Relaxation time: Linear trajectory	46
3.2	Relaxation time: Hyperbolic trajectory	49
3.3	Nonisolating versus Isolating integrals	54
4.1	The Milky Way according to Gaia (DR2)	72
4.2	Circular motion in the Galaxy	75
4.3	Epicyclic motion I	79
4.4	Epicyclic motion II	81
4.5	Vertical motion	84
4.6	Rotation curve	90
4.7	Miyamoto-Nagai model	91
5.1	Tully-Fisher relation	95
5.2	Fundamental plane	96
5.3	Rotational support of ellipticals	101
5.4	Spiral arm morphology	102
5.5	Spiral pitch angle	102
5.6	Lindblad resonances	104
5.7	Rotating patterns	105
6.1	G-dwarf problem	116

6.2 Mass-metallicity relation   118
6.3 Abundance ratio variations   119
7.1 Evolution of the density parameters   126
7.2 Density contrast   134
7.3 Spherical collapse   136
7.4 Transfer function   144
7.5 Cooling and the formation of galaxies   147
8.1 Stellar clusters   150
8.2 Steady tidal interaction   156
8.3 Tidal shock   158
8.4 Lowered Gaussian distribution   160
9.1 Two views of a galaxy cluster   164
9.2 Gravitational lensing effect   167
9.3 Mass profile from lensing   169
9.4 Environment and stellar population variations   172

List of tables

1.1 Typical photometric passbands   24
3.1 Relaxation times of typical stellar systems   48
7.1 Cosmological parameters   146

Preface

This book originates in a set of lectures I delivered at University College London between 2012 and 2018, corresponding to a master's degree module in the Physics and Astronomy Department. Although a good number of excellent published references on this material are available, my intended goal was to produce a set of notes that give a simplified, yet effective overview of the topic of galaxy formation and evolution, with special emphasis on dynamics. Extragalactic astrophysics started in earnest as a discipline of physics when galaxies were discovered as "island universes" in the 1920s. The first steps in this field focused on understanding the distance to the extragalactic nebulæ and to put them in context with the large-scale environment. Within the same decade, galaxies became very distant objects and unveiled the cosmological process of expansion, shaping our current view of the Universe. Further analysis concentrated on the details of stellar dynamics in galaxies and the physics driving the underlying components (dark matter, stellar populations, gas and dust). The advent of large galaxy surveys such as the Sloan Digital Sky Survey, and exquisite observations from facilities such as the Hubble Space Telescope have considerably transformed the field, allowing us to probe the distribution of galaxies over large cosmological scales, and to look at galaxy formation mechanisms 'under the microscope'. The material presented in this book provides an introduction to the field for advanced undergraduates and beginning postgraduates. No substantial background in astrophysics is expected, but good knowledge of calculus is needed to enjoy the physics of galaxies at its fullest.

I would like to thank the staff at UCL Press, especially Chris Penfold, for offering the opportunity to publish these lecture notes in an open access format. I would like to thank the previous lecturers of a precursor module to the one I taught, Jonathan Rawlings, Jeremy Yates and Mark Cropper, for outlining and developing such an exciting course. I also thank Anna Pasquali, Prasenijt Saha, Witold Maciejewski, Andrew Hopkins and Roger Davies for their support in putting together this book, and Joe Silk and Ofer Lahav for their guidance throughout the years. My interactive audience, the students who participated in this module, are warmly thanked for their input, and often inquisitive minds. I am especially grateful to Jennifer Chan, Lorne Whiteway and Ellis Owen. My wife, Isabel, will always have my immense thanks and gratitude for her undying support during the many weekends when these lecture notes were put together.

Horsham, West Sussex, July 2018

Acknowledgements

The images of M51 (figure 5.6) and M13 (figure 8.1) are based on photographic data of the National Geographic Society – Palomar Observatory Sky Survey (NGS-POSS) obtained using the Oschin Telescope on Palomar Mountain. The NGS-POSS was funded by a grant from the National Geographic Society to the California Institute of Technology. The plates were processed into the present compressed digital form with their permission. The image of M6 in figure 8.3 is based on photographic data obtained using the UK Schmidt Telescope. The UK Schmidt Telescope was operated by the Royal Observatory Edinburgh, with funding from the UK Science and Engineering Research Council, until June 1988 and thereafter by the Anglo-Australian Observatory. Original plate material is copyright © the Royal Observatory Edinburgh and the Anglo-Australian Observatory. The plates were processed into the present compressed digital form with their permission. The Digitized Sky Survey was produced at the Space Telescope Science Institute under US Government grant NAG W-2166.

The sketch of the Hubble fork diagram (figure 1.3) was created by modifying a set of images from the Sloan Digital Sky Survey (http://www.sdss.org).

It is a pleasure to acknowledge the National Aeronautics and Space Administration (NASA) and European Space Agency (ESA) for use of their superb and inspiring images from missions such as the Hubble Space Telescope and Gaia. The National Radio Astronomy Observatory (NRAO) is thanked for use of the radio image of M81 in figure 8.1 (rightmost panel).

We are grateful to *Astronomy & Astrophysics* and Institute of Physics (IoP; publisher of the *Astrophysical Journal*) for their straightforward reprint permission process. All authors cited in figures copied or based on ApJ, A&A, MNRAS and PASJ papers are warmly thanked for allowing us to use their work. Antony Lewis is thanked for the use of the Python code PᴙCAMB in figures 7.2 and 7.4. This code can be found at https://camb.info. Regarding figures from data published in MNRAS (figures 1.4, 5.2, 5.3, 6.1, 6.3, 9.3, 9.4): these are by permission of Oxford University Press on behalf of the Royal Astronomical Society. The material reproduced from the articles cited in the figure captions is not covered by the CC BY-NC-ND license of this publication. For permissions, please email journals.permissions@oup.com.

1
An introduction to galaxy formation

Galaxies are the building blocks of the Cosmos. Separated by vast distances, they also serve as tracers of the cosmic expansion and the primordial density fluctuations that gave rise to structure in the universe. Galaxy formation requires an understanding of the most fundamental physical processes: gravitation, statistical mechanics, gas hydrodynamics, radiative transfer, atomic physics, etc. In this book we will focus on the gravitational side of galaxies, dealing with both the statistical treatment of galaxies as an N-body system evolving purely under gravitational forces and with the growth of galaxies from evolving density fluctuations in an expanding Universe. This introductory chapter presents an overview of the field, including the observables typically used to study galaxies, the mechanisms underpinning galaxy formation and the characteristic timescales involved.

1.1 The main ingredients of a galaxy

A galaxy is a complex system bound by gravity. In our current paradigm, the gravitational potential is dominated by dark matter, whose distribution is much more extended than the visible part, and forms a spheroidal halo. The ordinary matter – loosely called "baryonic matter" – is made up mostly of hydrogen and helium, in the form of stars, diffuse and clumpy gas, dust, planets, etc. Although the dark matter dominates the mass budget – with a contribution of around 85 per cent in mass of the total matter content – emission in the electromagnetic spectrum is provided only by the baryons, except for potential, but hard to find dark matter particle annihilation events. Therefore, there is a substantial difference between mass and light in galaxies.

The gaseous component provides fuel for star formation. A highly complex set of processes involving gas infall, turbulence, radiative transfer, feedback from star formation and magnetic fields plays a role in the physics of star formation (something we will leave aside in this textbook). In addition, dust provides an important tracer of star formation as it is typically found in gas-rich star-forming environments. The scattering, absorption and emission of radiation from dust makes this component key in the thermodynamics of star formation. Galaxies with very high star formation rates (starbursts) are often enshrouded in dust, with the most active regions being practically opaque to optical radiation and displaying prominent emission in the infrared by heated dust: this is the case with Ultra-Luminous Infrared Galaxies (ULIRGs) or submillimetre galaxies (SMGs).

In addition to these components, it is worth noting the presence of a supermassive black hole (SMBH) at the centres of galaxies. With masses between a few million and several billion Suns, SMBHs can regulate the formation of their host galaxy. As gas accretes onto the SMBHs, a very luminous Active Galactic Nucleus (AGN) is formed. The energetic output from the AGN in the form of jets can affect star formation over the full scale of the galaxy, in ways that are still open to debate.

1.2 Observables

This section gives a nonexhaustive overview of the type of observables commonly applied to the study of galaxies.

Colours

In astrophysics, colour is defined as the flux ratio measured through different filters (see section 1.5). The interpretation of a colour depends on the wavelengths covered by the filters. In the ultraviolet/optical/infrared spectral windows, colour can be considered a rough proxy of stellar age. Light from younger stellar populations is predominantly contributed by massive, luminous, blue stars. However, other factors – such as chemical composition or dust – will affect this interpretation: a red colour need not imply old stars. For instance, the red colours found in so-called ERO galaxies (Extremely Red Objects) often originate from a young, but dusty stellar population. Figure 1.1 shows a mosaic of images of the nearby spiral galaxy M81, illustrating how a coverage of different regions of the electromagnetic spectrum allows us to study different processes in galaxies.

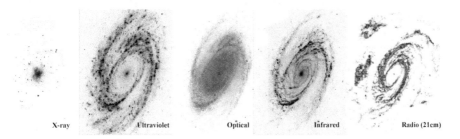

| X-ray | Ultraviolet | Optical | Infrared | Radio (21cm) |

Figure 1.1 Different views of nearby galaxy M81 (NGC3031, distance 3.7 Mpc). From left to right, images in the X-ray (*NASA/CXC/Wisconsin/ Pooley & CfA/ Zezas*), ultraviolet (*NASA/JPL-Caltech/CfA/Huchra et al.*), optical (*NASA/ESA/CfA/Zezas*), infrared (*NASA/JPL-Caltech/CfA*) and radio (*NRAO/AUI/Adler & Westpfahl*) spectral windows.

The X-ray image reveals a diffuse component tracing hot gas; a central bright source betraying the presence of a supermassive black hole; and a number of point sources that correspond to X-ray binaries – stellar systems where one of the members is a compact object (neutron star or black hole), whose strong gravitational potential drags and heats up the outer layers of the companion star. In contrast, the ultraviolet emission is due mainly to massive, young stars and reveals the sites of ongoing star formation. The optical and near infrared windows are dominated by the bulk of the stellar populations, whereas at longer wavelengths, in the far infrared, emission is produced by dust, that – like UV light – also traces sites of ongoing star formation. At even longer wavelengths, in the radio, emission originates from supernova remnants and HII regions (ionized hydrogen around star-forming sites), and at $\lambda = 21$ cm, we find resonant emission from neutral atomic hydrogen (HI).

Spectroscopy

The spectrum of a galaxy is the observed flux density as a function of wavelength, $F(\lambda)$, or frequency, $F(\nu)$. Galaxy spectra carry valuable information about the kinematics and the chemical composition of the stellar and gaseous components. Motions along the line of sight towards the observer affect the position and shape of the spectral features (both in absorption and emission) via the Doppler shift. For instance, the absorption lines of a massive galaxy are significantly broader with respect to the same features in a low-mass galaxy, an effect caused by the higher velocity dispersion of the stellar component. The bulk rotation of disc galaxies

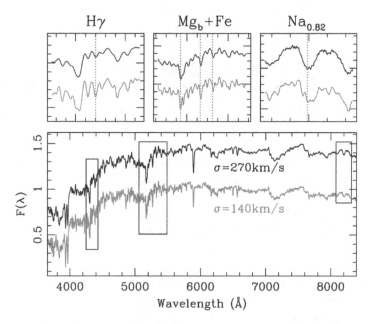

Figure 1.2 Extragalactic spectroscopy at work: high quality stacked spectra from the Sloan Digital Sky Survey are shown for two early-type galaxies with different velocity dispersion – broadly tracing the mass of the galaxy. The insets zoom in special windows that feature absorption lines sensitive to the age, chemical composition and mass distribution of the underlying stellar component. (Source: data from Ferreras et al., 2013, MNRAS, 429, L15.)

is measured by the characteristic 'S'-shaped pattern of the spectral line centres with respect to galactocentric distance. Moreover, a non-Gaussian analysis of the kinematic kernel – via higher order moments or a Gauss-Hermite expansion – allows us to further constrain the motion of the gas and stars of a galaxy.

The absorption lines in ultraviolet/optical/infrared galaxy spectra originate in the atmospheres of their stars. Therefore, they carry information about the properties of the stellar populations (see an example in figure 1.2). For instance, spectral line strengths such as Balmer absorption ($H\beta$, $H\gamma$, $H\delta$), or metallicity-dependent features such as the Mgb-Fe complex in the 5100-5400Å region provide constraints on the age and the chemical composition of galaxies. In emission, spectral lines originate from ionized regions, and the line luminosities constrain the properties of the gas, including its composition, temperature and ionisation state. For instance, the BPT diagram[1] (named after Baldwin, Phillips and

Terlevich) provides a simple diagnostic to discriminate between emission from star-forming and AGN activity, by comparing the ratio between two pairs of line luminosities, such as Hβ/[OIII] and Hα/[NII].

The dust component can also be probed with spectroscopy, including its overall attenuation effect with respect to wavelength and the presence of spectral features, most notably the NUV 2175Å "bump" or the silicate features in the infrared at 9.8 and 18 μm. Also in the infrared window, emission lines from polyaromatic hydrocarbons (PAH) and thermal radiation are also sources of information regarding dust.

Surface brightness

Surface brightness (SB) is defined as the flux received from a section of the galaxy, i.e., within a solid angle. Therefore, it is not possible to measure the surface brightness of an unresolved source (e.g., the vast majority of stars observed with standard techniques). A simplified characterization involves the definition of elliptical isophotes, regions with the same surface brightness, leading to a one-dimensional surface brightness profile $\Sigma(\theta)$, where θ is the angular galactocentric distance, measured, e.g., as the semi-major axis of the ellipse describing the isophote. This distance can be translated into the physical projected two-dimensional, radius R, by use of the (angular diameter) distance (D_a): $R = D_a \tan\theta$ (note R is often defined as a circularized radius $\sqrt{ab}$, where a and b are the semi-major and semi-minor axis, respectively, of the corresponding ellipse that describes the galaxy). Although trivial for nearby sources, there is a second measure of the distance that translates between luminosity (L) and flux (F). This so-called luminosity distance (D_l) is defined such that $F = L/4\pi D_l^2$. For large but noncosmological separations (i.e., where the general relativistic effects are unimportant), D_a and D_l are practically the same, making the surface brightness independent of distance, but for cosmological separations $D_l = (1+z)^2 D_a$, where z is the redshift to the source (see chapter 7). The radial distribution of the surface brightness of disc galaxies can be described by an exponentially decaying profile:

$$\Sigma(R) = \Sigma_h e^{1-R/h},\tag{1.1}$$

where h is the disc scale length, and Σ_h is the SB at $R = h$. In contrast, elliptical galaxies have a much steeper profile (de Vaucouleurs, or $R^{1/4}$ profile):

$$\Sigma(R) = \Sigma_e e^{7.67\left[1-\left(\frac{R}{R_e}\right)^{1/4}\right]}.\tag{1.2}$$

The effective radius R_e is defined such that half of the total flux is enclosed within R_e, and Σ_e is the surface brightness at R_e. A generic expression often used in the description of the SB distribution is the Sérsic profile:

$$\Sigma(R) = \Sigma_e e^{-\kappa\left[\left(\frac{R}{R_e}\right)^{1/n} - 1\right]}, \tag{1.3}$$

where n is the Sérsic index, and $\kappa = 1.9992n - 0.3271$ is a normalization factor that ensures the flux within $R \leq R_e$ is one half of the total flux.[2] This profile includes the exponential case for $n = 1$ and the de Vaucouleurs profile for $n = 4$. The value of the index is commonly used as a quantitative indicator of morphology, with the early-type galaxies having $n \gtrsim 2.5$, and late-type galaxies having lower values of n.

Morphology

The morphology of a galaxy gives us information about the distribution of stars, gas and dust. Note that the appearance of a galaxy strongly depends on the spectral window (see figure 1.1). Morphological studies normally concern the stellar component as the main tracer of the gravitational potential of the galaxy. Galaxies are split roughly into three main groups: elliptical, spiral and irregular. The presence of a bar in spiral galaxies motivates a branching in the classification scheme, illustrated by the Hubble tuning fork diagram (figure 1.3). Elliptical and lenticular galaxies are combined into early-type galaxies and present a spheroidal distribution, explained by a major merging event, or some collapse mechanism by which the total angular momentum was kept low. The oblateness of these systems cannot be fully explained by rotation (see chapter 5), and their spectral energy distribution corresponds to old and metal-rich stellar populations, which reflect an early, intense and efficient process of star formation and are corroborated by their chemical composition (see chapter 6).

In contrast, spiral galaxies, more aptly described as disc (or late-type) galaxies, are flatter systems where a large fraction of the total kinetic energy is in the form of bulk rotation. For example, the (thin) disc of our Milky Way galaxy has a vertical extent about one-tenth of the disc size. The collapse of gas under gravity can develop such a rotating structure, as for instance, during the formation of our solar system. Disc galaxies have a more complex distribution of stellar populations than ellipticals, featuring ongoing star formation as well as a substantial presence of old stars. The central part of a disc galaxy usually hosts a spheroidal structure,

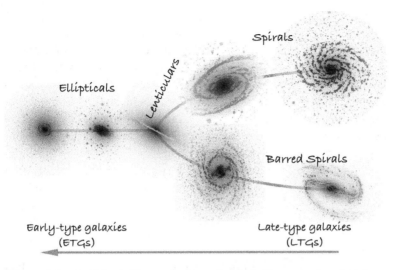

Spirals

Ellipticals

Lenticulars

Barred Spirals

Early-type galaxies
(ETGs)

Late-type galaxies
(LTGs)

Figure 1.3 Hubble-de Vaucouleurs tuning fork diagram, showing the major morphological classification of galaxies into ellipticals (*left*) and spirals (*right*), the latter consisting of standard and barred spiral galaxies. (Source: The images of the galaxies were created from observations taken by the Sloan Digital Sky Survey.)

a bulge, whose origin also constitutes an open problem, with some bulges resembling an early-type galaxy (classic bulges) and others being the product of secular dynamical evolution (pseudo-bulges). Spiral arms are the most conspicuous features of disc galaxies. Their origin is based on dynamical resonances, which will be briefly explored in chapter 5. Irregular galaxies are more complex dynamical structures, often betraying the presence of an ongoing merger or tidal interaction with a neighbour.

The standard method of morphological analysis involves visual inspection of images, preferably done through several filters. Morphology can also be determined from the surface brightness profile (see above), or by the application of alternative methods involving nonparametric observables, which do not make any assumption about the radial surface brightness profile. A number of observables are defined such as the concentration, asymmetry or clumpiness, or even higher order moments of the pixellated surface brightness distribution.[3] More recently, machine learning methods are being applied to perform "visual classification" in a fully automated way, using, for instance, artificial neural networks trained on visually classified data sets.[4]

Size

It is difficult to provide a clear-cut definition of the size of galaxies. Being diffuse objects, their borders are fuzzy. One typically simplifies the problem to a one-dimensional equivalent by binning the observation radially within elliptical isophotes. A traditional definition of size, D_{25}, hypothesizes that the galaxy extends in a region brighter than 25 mag arcsec^{-2} in the B band. This choice is motivated by the fact that surface brightness does not vary with distance over noncosmological scales. However, we know that over large distances, the surface brightness does indeed change, decreasing as $(1+z)^4$, where z is the redshift (termed 'Tolman dimming'). A more robust criterion is based on the Petrosian radius, derived from the following expression:

$$\eta(R_0) \equiv \frac{\Sigma(R_0)}{\langle \Sigma(R) \rangle_{R<R_0}}.$$ (1.4)

This function starts at $\eta \sim 1$ for $R_0 \sim 0$ and decreases outwards. The Petrosian radius is defined as the value R_P for which $\eta(R_P) = 0.2$. Its being a ratio of surface brightness eliminates the dependence on the cosmological Tolman dimming. Another option involves fitting the observed data with a generic surface brightness profile such as the Sérsic function presented above (equation 1.3), so that the parametric effective radius of the function can be used as a measure of size.

Exercise 1.1

Find the Petrosian radius as a function of h for the exponentially decaying surface brightness case shown in equation 1.1.

Luminosity function

Galaxies can be classified according to their absolute luminosity (i.e., their power, or energy emitted per unit time, usually defined with respect to solar luminosity, $L_\odot$, or given by their apparent flux when located at a fiducial distance). The volume number density of galaxies (n) per luminosity interval is the Luminosity Function. It can be suitably described by a power law with an exponential cutoff at the bright end, defined as the Schechter function:

$$\frac{dn}{dL}dL \equiv \Phi(L)dL = \Phi_0\left(\frac{L}{L_\star}\right)^\alpha e^{-L/L_\star}\frac{dL}{L_\star}, \qquad (1.5)$$

with three free parameters: the characteristic luminosity ($L_\star$, usually given in units of $L_\odot$, or as an absolute magnitude, $M_\star$), the power law index, equivalent to the slope at the faint end (α), and the normalisation, given by the number density of $L_\star$ galaxies (given, aside from a factor e, by Φ_0, usually in units of Mpc^{-3}). We will see in chapter 7 that the shape of the Schechter function is a direct consequence of galaxies being created from random density fluctuations following collapse once the fluctuations are large enough.

The parameters describing the luminosity function change with respect to galaxy morphology. For instance, elliptical galaxies are less numerous (lower Φ_0), but are brighter (higher $L_\star$) than disc galaxies. Moreover, these populations change with respect to environment, with ellipticals being more prevalent in galaxy clusters (a property often termed the 'morphology-density relation'). An alternative description of the galaxy census is the stellar mass function, with a similar notation as in the Schechter function, but referring to mass instead of luminosity. We will see below that a comparison between the stellar mass function and the dark matter mass function expected from a simple theoretical argument gives a very powerful diagnostic of the nontrivial relation between dark matter halos and galaxies.

Exercise 1.2

Show that if we adopt the Schechter luminosity function (equation 1.5), we need $\alpha > -2$ if the total luminosity (integrated throughout the whole galaxy population) is finite. Then consider the $\alpha = -1$ case: find the total luminosity from all galaxies, and show that the total number of galaxies diverges. How can an infinite number of galaxies have a finite total luminosity?

Star formation rate

The star formation rate (SFR, often denoted by the greek letter ψ) is defined as the mass content in stars created per unit time. This can be defined over different scales as an integrated quantity over a cosmological volume, for a single galaxy, or as a volume (or projected surface) density of star formation within a galaxy. The average star formation rate of the Milky Way galaxy at present is $\psi_{MW} \sim 1 M_\odot \, yr^{-1}$, and star-bursting galaxies can sustain rates hundreds (or even thousands) of times higher. Measurements of the SFR use indirect tracers, focusing either on the presence of young stars or on the fuel that triggers formation. The ultraviolet emission from a galaxy reveals the presence of massive (therefore young) stars. Emission lines from ionized gas also betray the presence of the most massive stars, hot enough to ionize large volumes surrounding the star-forming regions. The (hydrogen) gas can be observed both in atomic and molecular form (the latter normally detected indirectly through the presence of CO), and the dust – which is abundant in star-forming regions – can also be probed by attenuation of starlight or by emission at infrared wavelengths. There is a whole industry of extracting star formation rates from observations, beyond the scope of these lecture notes.[5]

Environment

'Environment' refers to the large-scale regions where galaxies are found. It can be characterized by a density averaged over volumes that include many galaxies. It provides a link between local formation processes in galaxies and the large-scale structure of the Universe (figure 1.4). Roughly four different environments can be considered: voids (underdense regions, away from large-scale filaments), field (representing average densities over cosmological volumes), groups (gravitationally bound structures that include several massive galaxies, with velocity dispersions among group members comparable with the velocity dispersion of stars within galaxies, i.e., $100–300 \, km \, s^{-1}$) and clusters (the highest density regions, including thousands of galaxies within a $\sim 1 \, Mpc$ radius, and velocity dispersions $\gtrsim 500 \, km \, s^{-1}$; see chapter 9). Another classification of environment can be given by the mass of the dark matter halo that hosts a given galaxy. Moreover, inside groups and clusters, the environment of the central region is very different from the cluster outskirts, and the distinction between a central (i.e., the most massive galaxy in a dark matter halo) and a satellite appears to be fundamental in the description of galaxy formation.[6]

There are important environment-related processes that affect the evolution of galaxies: mergers, strangulation, ram pressure stripping, harassment. The environment is tightly linked with the initial conditions of the system: clusters are structures formed from higher (and rarer) fluctuations of the primordial density distribution. Furthermore, because of the inherent growth of structures with cosmic time (bottom-up formation), clusters can be found only at relatively later cosmic epochs.

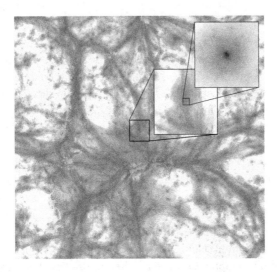

Figure 1.4 A slice of the Universe as seen by the EAGLE galaxy formation project. The cosmic web is readily apparent, with filaments, voids and clusters (at the confluence of filaments). The zoomed-in panels give an idea of the size of a typical galaxy within the cosmic web. Environment-related processes will shape the formation histories of galaxies. (Source: Schaye et al., 2015, MNRAS, 446, 521.) Used by permission of Oxford University Press.

Nuclear activity

Most galaxies harbour a supermassive black hole (SMBH) at their centres, with a mass in the range 10^6–$10^9 M_\odot$. When gas is available, as it flows into the SMBH, it forms an accretion disc where the material gets heated, emitting copious amounts of radiation across a wide range of wavelengths and producing an AGN that can outshine the whole galaxy. Only a few galaxies display AGN activity, but it is believed that most massive galaxies experienced AGN activity some time during their past formation history. There are many types of AGN: Quasars, Seyferts, Radio Galaxies, Liners, BL Lacs, etc.

The fundamental issue related to galaxy formation is that AGN activity can affect the flow and heating of the gas that fuels star formation in galaxies. Therefore, although AGNs are confined to a minuscule volume within the galaxy (a $10^9 M_\odot$ SMBH has a radius $r_g = 2GM/c^2 \approx 20$ AU and a region of influence $r_i = GM_\bullet/\sigma^2 \approx 50$ pc), they can control galaxy formation (i.e., over scales $\gtrsim 50$ kpc). It is believed that AGN activity provides a feedback mechanism that contributes to the quenching of star formation in massive galaxies. A strong correlation is found between galaxy properties and the mass of the central black hole.[7]

Distance

A simple but fundamental observable is the distance to galaxies from our vantage point. Measuring distances beyond the solar system entails a progressive set of steps (called the 'cosmological distance ladder') to measure successively larger distances. The first 'extragalactic' rung of the ladder is based on the use of variable stars, especially bright Cepheids. Their period-luminosity relation allows us to determine their true luminosity (L) from a sequence of observations with time (the light curve). In combination with the observed flux (Φ), the distance (d) is obtained via $\Phi = L/4\pi d^2$. This method pioneered the field of extragalactic astrophysics, when our neighbour, the Andromeda galaxy, was found to lie much farther out than any of the visible stars. Additional steps of the ladder involve the use of tight scaling relations among galaxies, such as the Tully-Fisher relation (chapter 5). Closer to home, the first step of the ladder uses basic trigonometry: by carefully measuring the positions of stars, we find that in addition to their proper motion (caused by their relative velocity with respect to the Sun), they follow a conspicuous yearly cycle, with all stars moving "in phase" but with different amplitudes. This motion is caused by parallax, the apparent change in the observed position as the Earth orbits the Sun. Therefore, those with the largest cycles are closest to us. These cycles are nevertheless very small, and thus limit useful estimates of parallax to the nearest stars.[8] A natural unit of distance is therefore the one at which a star would have a parallax of 1 arcsec. This is called the 'parsec' (pc), and trivially amounts to $\sim 1\,\mathrm{AU}/\pi$, where π is the parallax in radians and AU denotes an astronomical unit (the average Sun-Earth separation, about 150 million kilometres). Therefore, 1 pc amounts to 3.086×10^{16} m, or 3.26 light-years. Distances within galaxies are measured in kiloparsec, and galaxy-galaxy separations are measured in Megaparsec (Mpc) or Gigaparsec (Gpc). The distance from the Sun to the centre of the Milky Way galaxy is approximately 8 kpc, and the distance to the neighbouring Andromeda galaxy is 0.8 Mpc.

Redshift

At the top of the cosmological distance ladder, the separation between galaxies takes advantage of redshift. Due to the expansion of the Universe, photons with wavelength λ_0, emitted from a distant galaxy (at cosmic time t), are observed (at time t_0), at a longer wavelength, λ_{obs}, stretched by a factor z, termed 'redshift':

$$\frac{\lambda_{\text{obs}}}{\lambda_0} \equiv 1 + z = \frac{a(t_0)}{a(t)}, \qquad (1.6)$$

where z is the redshift, and $a(t)$ is the scale factor of the Universe, a dimensionless quantity that represents the global cosmic expansion. The scale factor at present time, $a(t_0)$ can be arbitrarily set to one. Since redshift increases towards earlier times, it directly correlates with distance. Note that in groups of nearby galaxies the relative motions (termed 'peculiar velocities') will also affect, via the Doppler effect, the observed redshift. Only at very large distances does the cosmological effect overwhelm any contribution from peculiar velocities. We will see in chapter 7 that it is convenient to separate the uniform expansion (termed the 'Hubble flow') and the peculiar motion, by use of co-moving distances between galaxies where the contribution from the uniform expansion is factored out. We will also see in that chapter that the concept of distance becomes less trivial, and different distances to the same object will be obtained when using alternative definitions (section 7.1). Measuring galaxy properties over a range of redshifts implies testing the evolution of galaxies across cosmic time. As reference, at redshift $z=2$, galaxies are found in a Universe that is only 3 Gyr old, i.e., about 20 per cent of its current age. At very high redshift, cosmic age poses a stringent constraint regarding the timescales expected in the formation of galaxies. For instance, bright quasars or galaxies at $z \sim 10$ have only about 500 Myr of time to evolve, so observations at such high redshift substantially reduce the number of potential scenarios of formation.

1.3 Physical processes

We switch from the observational properties of galaxies to the physical mechanisms of galaxy formation. This section gives a succinct view of some of the processes that play an important role in the formation and evolution of galaxies.

Initial conditions

Galaxies develop from small density perturbations in the dark matter distribution at early times. Under gravity, these perturbations grow within an expanding Universe, until virialisation is achieved, forming stable structures called 'dark matter halos'. In the current cosmological framework, about one-sixth of all matter is in the form of baryons (mostly

H and He). Originally in gaseous form, this material dissipates energy –
in contrast with the collisionless dark matter. The gas falls to the centre of
the halo, cooling down and eventually forming stars.

Certain initial conditions in the overdensities are needed to explain the properties of galaxies: too low and there will not be enough
galaxies at present; too high, and we would be surrounded by a large number of massive, old galaxies, at odds with the observational evidence.
The present distribution of galaxies is compatible with a model where
~24 per cent of the total matter/energy content of the Universe is in
the form of dark matter, an additional ~4 per cent is in the form of baryons and the remaining ~72 per cent is made up of an unknown energy
field commonly known as 'dark energy', that produces an accelerated expansion (this is the standard ΛCDM framework, where Λ refers to the
dark energy term, and CDM stands for cold dark matter). Galaxy formation is inherently tied to cosmology, and moreover to the fundamental
properties of particles. In fact, many cosmology-oriented experiments
use large galaxy surveys to map the large-scale structure of the Universe
(e.g., the Sloan Digital Sky Survey, the Dark Energy Survey, or the Euclid
mission).

Gravitational instability

Small fluctuations are rapidly amplified by gravity – as neither expansion
nor the repulsive effect of dark energy are effective over galaxy scales.
It is useful to quantify the evolution of the fluctuations with the density
contrast, defined as follows:

$$\delta(\vec{r};t) \equiv \frac{\rho(\vec{r};t)}{\langle\rho\rangle(t)} - 1. \tag{1.7}$$

In a noncosmological environment, where the effects of expansion are
negligible (e.g., a gas cloud within the Milky Way), a small overdensity
will grow exponentially with time ($\delta \propto e^{t}$). Over cosmological scales, the
expansion will slow down this growth rate to a power law of cosmic time,
$\delta \propto t^{\alpha}$, with the power law index depending precisely on the expansion
rate, given by the cosmological model. Cosmologists often work in the
linear regime ($\delta \ll 1$), where the growth equations can be expanded as a
Taylor series, and truncated to the lowest order (i.e., linear) terms. In contrast, astrophysicists tend to work instead in the highly nonlinear regime
($\delta \gg 1$), where galaxies constitute huge overdensities with respect to the
background density of the Universe.

The hierarchical growth of structure inherent to the current cosmological paradigm implies that fluctuations over small scales are the first to form virialized, stable systems at early times, whereas the more massive structures assemble later. Therefore, the linear regime corresponds at present time to very large scales (of cluster/supercluster size), and the equivalent linear regime for galaxy scales took place at earlier times. Hence, cosmology studies using galaxies need to probe very large volumes. Over smaller scales (galaxy clusters or groups), galaxy-related processes overwhelm any imprint on the observations related to cosmology.

Gas cooling and star formation

The only channel of interaction among dark-matter particles is via gravity, in a collisionless way, i.e., simply following the trajectories dictated by the overall gravitational potential. In contrast, gas particles (hydrogen/helium) can lose energy, as they also interact through electromagnetic forces. Photons emitted from collisions of gas particles are lost from the system, and act as an energy sink. The cooling rate depends on the physical properties of the gas: temperature, density, composition. At very high temperatures ($T \gtrsim 10^7$ K), the gas is fully ionized, and cools mainly through bremsstrahlung emission from free electrons. At lower temperatures (10^4 K $< T < 10^7$ K) atomic recombination/collisional excitation processes dominate, in ways that are strongly dependent on the species present (i.e., the metallicity of the gas). At even lower temperatures ($T < 10^4$ K), the gas is almost completely neutral, and collisional excitation of fine structure energy levels (atoms/ions) and rotational/vibrational energy levels (molecules) will contribute significantly in this regime. We will see in chapter 7 that the balance between gas cooling and the dynamical timescale from gravity is responsible for the characteristic sizes and masses of galaxies. To quantify the role of cooling, it is convenient to define a cooling timescale as the time it would take to remove the internal energy of the gas at the rate dictated by its cooling. It is proportional to ($T/n\Lambda$), where T is the temperature of the gas, n is its number density, and Λ is the cooling rate, a complex function that, in its simplest form, depends on temperature and chemical composition.

The cooling of atomic gas leads to pressures and temperatures at which molecular gas can form. Following the formation of molecular clouds, fragmentation of clumps create the so-called pre-stellar cores, which act as seeds of star formation. This process poses one of the most difficult and challenging problems in physics. For instance, the distribution of stellar masses in a newly forming region (called the 'Initial Mass Function', see chapter 6) remains an open problem as analytic models cannot describe the complexity of the underlying processes, and hydrodynamical simulations lack the necessary dynamical range to probe these processes in detail.

From a phenomenological point of view, we make use of scaling relations backed by the observations that allow us to describe in simple terms the process of star formation in galaxies. One of the most fundamental scaling relations is the one between the star formation rate (SFR) and the gas content. The SFR is found to follow a power law of either the volume gas density (Schmidt law) or the surface gas density (Kennicutt law). For instance, the latter is defined as follows:

$$\frac{\Sigma_\psi}{\mathrm{M_\odot yr^{-1} kpc^{-2}}} = 1.6 \times 10^{-4} \left(\frac{\Sigma_{\mathrm{gas}}}{1 \mathrm{M_\odot pc^{-2}}}\right)^{1.4}. \tag{1.8}$$

The power law index can be derived from a straightforward theoretical argument: $\psi \propto \rho_{\mathrm{gas}}/t_{dyn}$, where $t_{dyn} \propto \rho_{\mathrm{gas}}^{-1/2}$ is the dynamical timescale of collapse of the gas (see exercise 1.3). More recent observations suggest this law is valid when the densities correspond to the molecular (not atomic) gas.[9]

Figure 1.5 illustrates the overall link between the mass of a galaxy and its formation history. The left panel shows the colour-mass diagram of low redshift galaxies – colour can be used as a rough proxy for age, with red colours ($g - r \gtrsim 0.8$) meaning old populations. At the massive end, most galaxies – at late cosmological times – populate an elongated region towards the red side (top of the figure) called the 'red sequence', consisting of old stars. These are galaxies whose stellar component was formed at early times. In contrast, low-mass galaxies are markedly blue, populating a region called the 'blue cloud' and representing star-forming systems. The panels on the right give a very schematic view of the interplay between the red sequence and the blue cloud. Mergers will induce mass growth in galaxies. The ages of the stellar component reveal the age of the progenitors that merge into a more massive system, as well as potential star formation, when the merger includes gas (black arrows, termed 'wet mergers'). Mergers between galaxies that have no gas (white arrows, 'dry mergers') mix their stellar content without adding any younger component. However, this picture is, alas, too simplistic. Incidentally, the region between the blue cloud and the red sequence is called the 'green valley', and it may hold valuable clues about galaxy evolution processes, as it represents a transition stage from star-forming galaxies to quiescent

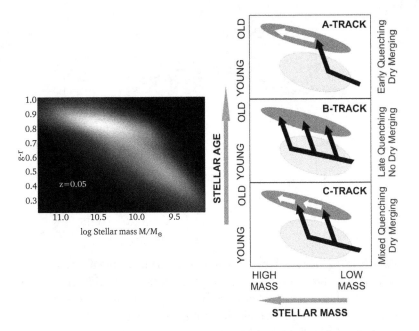

Figure 1.5 Evolution of galaxies on the colour versus mass diagram. The left panel shows the *observed* distribution of galaxies from the Sloan Digital Sky Survey (https://www.sdss.org), coding the number density of galaxies as a grey scale. The panels on the right give a number of schematic trends of evolution, with black (white) arrows representing gas-rich (gas-poor) mergers. (Source: adapted from Faber et al., 2007, ApJ, 665, 265.)

systems. It is in the green valley where we expect the feedback-driven processes that lead quenching to operate.

Feedback

The simple laws that relate the star formation activity in a galaxy to the amount of gas – such as the Kennicutt law of equation 1.8 – also depend on additional properties that can be encoded as a star formation efficiency. For instance, in its basic form, Schmidt's law ($\psi = \nu\rho_{gas}^{1.5}$) leaves a free parameter (ν) that relates to the efficiency of the process. Various mechanisms can reduce the star formation rate, even if gas is present. The most direct one is star formation itself: massive stars end their lives as supernova explosions, releasing vast amounts of energy both as a heat source and as ejected material ploughing through the gaseous

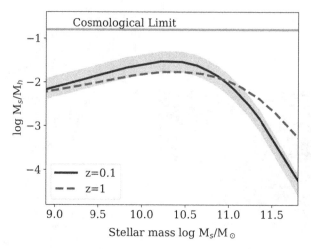

Figure 1.6 The ratio between galaxy (stellar) mass and dark matter halo mass is shown as a function of galaxy mass, at two different redshifts. The top horizontal line is the cosmological limit ($\approx \frac{1}{6}$) derived from the Planck 2015 cosmological parameters (see table 7.1). (Source: adapted from Behroozi et al., 2010, ApJ, 717, 379.)

component that feeds star formation. Since star formation is a highly clustered process, many supernova explosions can trigger a super-wind that will drag substantial fractions of the gas into intergalactic space, in a form of negative feedback (termed 'stellar feedback'). A second feedback process involves the activity from a central Active Galactic Nucleus, ejecting gas and heating the surrounding regions, in a similar way to the above, and suppressing star formation.

One of the key figures that conveys the most fundamental issues of galaxy formation (figure 1.6) compares the mass function of dark matter halos (from numerical simulations of structure formation) with the observed stellar mass function of galaxies. Making the assumption that one can match the abundance of dark matter halos to galaxies (the former being hosts to the latter), it is possible to define the mass ratio between dark matter halos and stellar mass in galaxies (this procedure is called 'abundance matching'). The figure shows that at both ends (massive galaxies on the right and dwarf galaxies on the left), the mass ratio decreases, suggesting that a mechanism must be present to reduce the efficiency of conversion from gas into stars. Stellar feedback is believed to control the low-mass end, producing an increasing correlation of the ratio with galaxy mass: lower mass galaxies will have shallower gravitational potentials, ejecting more gas, and therefore reducing the star formation rate. In contrast, massive galaxies have stronger potential wells, making stellar feedback less effective. AGN-driven feedback is supposed to drive the negative correlation at the high-mass end, as the mass of the SMBH – and therefore the energetic output from the AGN – will increase with galaxy mass. Also note that at the peak of the trend, corresponding to galaxies with stellar masses $\sim 3 \times 10^{10} M_\odot$, the ratio is about 4 per cent, substantially lower than the 'primordial' ratio between baryons and dark matter (~ 16 per cent, top horizontal line). Moreover, galaxies at the peak of the ratio do not have a vast amount of gas (or dust) to compensate for this mismatch, suggesting that only a small fraction of the available baryons is found in galaxies, with the remainder either not forming part of the galaxy formation process or being ejected during the main formation phases. The redshift evolution, from the dashed line (at $z = 1$) to the solid line (at $z = 0.1$) illustrates the significant growth in the dark matter halo at the massive end, without a comparative stellar mass growth – another sign of the mismatch between the growth of dark matter halos and galaxies.

Galaxy mergers and dynamical evolution

Dark matter halos (and their galaxies) interact among themselves, leading to a hierarchical growth process, whereby small structures – formed at early times – lead to more massive structures via mergers. When halos of a similar mass merge (major mergers), the gas in the progenitors is shock-heated to the new equilibrium (virial) temperature of the newly formed halo. Galaxies in the progenitor halos will also merge, producing a new galaxy at the centre of the newly formed halo. When the mass ratio of the halos is larger (minor merging), the merging may not be so effective, keeping the galaxy from the small halo as a satellite. Merging processes lead to morphological changes that strongly depend on the environment. For instance, spheroidal galaxies, believed to form from major merging processes, are more abundant in the higher density regions of clusters than in the field. Massive galaxies are more compact at high redshift, suggesting a substantial growth in size through galaxy mergers.

Internal dynamical effects may also play a role in the evolution of a galaxy. Instabilities can redistribute mass and angular momentum, changing the morphology of a galaxy. For instance, a thin disc with a high surface mass density is susceptible to instabilities that produce structures similar to those found in barred spiral galaxies. This process will efficiently drive gas towards the centre of the galaxy, fuelling star formation or AGN activity. Once formed, a bar can buckle away from the plane of the disc, to form a spheroidal object, called a 'pseudo-bulge',[10] which should be distinguished from classic bulges, themselves formed via mergers. Hence, the morphology of a galaxy can be driven either by external mechanisms such as the merger history or by internal (secular) processes.

Dynamical evolution introduces a characteristic timescale. In exercise 1.3 we find that the collapse time of a gravitating system with negligible internal kinetic energy is uniquely determined by its density (assuming no other forces oppose gravity). The dynamical timescale ($t_{dyn} \propto \rho^{-1/2}$) broadly introduces a lower limit on the duration of the 'settling' time that a galaxy undergoes after a dynamical event such as a merger or sudden mass loss.

Exercise 1.3

The dynamical timescale can be defined by the time it would take for a system driven only by gravity to collapse to a point (neglecting the initial contribution from kinetic energy). We assume each particle moves only under the gravitational potential of the collapsing mass, neglecting any nongravitational effects or local interactions with neighbouring particles. Show that this timescale depends only on the density of the galaxy as follows:

$$t_{dyn} = \sqrt{\frac{3\pi}{32G\rho}}.$$

Note that for the Sun ($\rho_\odot = 1.4\,\mathrm{g\,cm^{-3}}$), this time is approximately 30 min, i.e., the time it would take to collapse if it suddenly transformed into pressureless dust. This type of problem will reappear in chapter 7 with Friedmann's equations and the spherical collapse model.

Chemical enrichment

The 'primordial' periodic table of the elements is comprised only of hydrogen (76 per cent), helium (24 per cent) and traces of lithium. The phase of cosmological nucleosynthesis – when the temperature and density are appropriate for an efficient processing of nuclei – happened a few minutes after the Big Bang, but it was not long-lived enough to allow for the efficient creation of nuclei more massive than helium. The chemical composition of galaxies is determined, instead, by stellar nucleosynthesis. As stars evolve, new elements are synthesized in their hot and dense interiors via either thermonuclear reactions in the core, neutron capture, or explosive processes in the most massive stars. These elements are released during the final stages of stellar evolution, contributing to the change in composition (metallicity) of new generations of stars.

Hence, the evolution of the composition of stars with age (chemical enrichment) constitutes an important tracer of the formation history of a galaxy, as we will see in chapter 6. For instance, we can use the metallicity of stars as a rough proxy of age: extremely metal-poor stars are relics from the early formation phases of the Milky Way. However, the efficiency of star formation can alter this result: the cores of massive elliptical galaxies feature high metallicity, but they are also very old, reflecting a very intense process of formation that quickly enriched the interstellar medium at early times. The positive correlation between galaxy mass and its metallicity (measured both in the stellar populations and in the gaseous component) can be explained via feedback processes that trigger a metal-rich wind, preferentially removing material from the shallower gravitational potential wells of lower mass galaxies.

Evolution of stellar populations

The stellar populations in galaxies feature a wide range of stellar mass, age, and chemical composition. Observational data in the ultraviolet/ optical/infrared spectral window, covering approximately a wavelength interval from 0.1 to 4 micron, allow us to study the underlying stellar populations. Synthetic models combine our knowledge about stellar structure and evolution, producing spectra for a choice of parameters (age, metallicity, initial stellar mass distribution, etc.).[11] These models can be used to backtrack the star formation histories of galaxies. Although fraught with degeneracies inherent to the complexity of the population mixtures and the severe overlapping of spectral features on the stellar atmospheres, these models have allowed us to shed light on the connection between various observational properties and the star formation histories of galaxies. For instance, it has been found that stellar mass is the dominant driver of the overall formation history of elliptical galaxies, with massive galaxies featuring an earlier and more intense process of formation, whereas low-mass galaxies extend their formation over much longer timescales.[12]

1.4 Stellar clusters

Smaller than galaxies, stellar clusters are associations of stars with total stellar mass below $\sim 10^6 M_\odot$. They constitute sites of recent or past star formation – watch out for a notation overload, as we also use the term 'clusters' to refer to galaxy clusters! The population of stellar clusters is split into two main categories. (1) Globular clusters (GCs) are highly concentrated, spherical stellar systems whose trajectories do not follow the disc. Dominated by old stellar populations with relatively low metallicities, GCs are formed during an early and intense period of star formation. (2) Open clusters (OCs) are stellar associations found in the disc of the galaxy. They represent sites of recent or ongoing star formation, and their chemical composition is more evolved than that of GCs. As a stellar cluster ages, the system dissolves into the general distribution of disc stars via various mechanisms, induced mainly by tidal forces within the galaxy. OCs are less massive than globular clusters (with stellar masses around a few $\times 10^3 M_\odot$). GCs and OCs in our galaxy and nearby systems (most notably the Magellanic Clouds) can be observed in detail. In distant galaxies it is possible to observe GCs as unresolved sources, and they can be used as dynamical tracers of the underlying gravitational potential. Moreover, star-bursting galaxies feature super-star clusters, very massive and young stellar associations that can be considered a young version of GCs.

1.5 A technical note on astronomical observations

Flux density (or irradiance) is the main observable of an astronomical source. It is defined as the energy received per unit time and surface (as in, e.g., the collection area of a telescope). Typical units are $\mathrm{erg\, s^{-1}\, cm^{-2}}$ (cgs). However, flux densities are often quoted as magnitudes on a logarithmic scale:

$$m = Z - 2.5 \log F = Z - 2.5 \log \frac{L}{4\pi D^2}, \qquad (1.9)$$

where $Z = 2.5 \log F_0$ is the zero point, which corresponds to the flux of a reference star that, by definition, has zero magnitude. Traditionally, Vega (α Lyr), one of the bright stars in the (Northern) Summer triangle has been used as reference (see below). D is the distance to the source, and L is the luminosity (or power, energy emitted per unit time). In the above equation, m is the apparent magnitude. In order to compare the luminosities from sources located at difference distances, it is useful to factor out the distance by defining the absolute magnitude (M) as the apparent magnitude that the same source would have if located at a fixed, reference distance ($d_0 = 10 \, \text{pc}$). Since the flux decays as $1/D^2$, we have

$$M = m - 5 \log \frac{D}{d_0}. \tag{1.10}$$

The term $5 \log D/d_0$ is called the 'Distance Modulus'. This is a simple expression where redshift and other cosmology-related effects are negligible. So far we did not take into account the fact that a source emits photons over a range of wavelengths, according to some distribution (called the 'spectrum'). We define $F(\lambda)$ as the observed flux density per unit wavelength, also called the 'spectral irradiance' (measured in, e.g., $\text{erg} \, \text{s}^{-1} \, \text{cm}^{-2} \, \text{Å}^{-1}$). Alternatively, the spectrum can be given per unit frequency, $F(\nu)$; a standard unit is the Jansky (1 Jansky $= 10^{-23} \text{erg} \, \text{s}^{-1} \, \text{cm}^{-1} \, \text{Hz}^{-1}$). Note that $F(\nu) \neq F(\lambda)$, as one is measured per unit frequency and the other is measured per unit wavelength. We can relate both using $dF = F(\lambda)d\lambda = F(\nu)d\nu$, so that

$$F(\nu) = \frac{\lambda^2}{c} F(\lambda). \tag{1.11}$$

The total, or bolometric, flux consists of energy from all wavelengths, or frequencies: $F = \int F(\lambda)d\lambda = \int F(\nu)d\nu$. In spectroscopy, the observations are given as either $F(\lambda)$ or $F(\nu)$. Another expression often used to describe the spectrum is $\lambda F(\lambda)$, proportional to the number density of photons received on a detector per unit time. In photometry, we usually define the flux density passing through a filter with a transmission throughput $S(\lambda)$ as

$$F_S = \frac{\int_0^\infty F(\lambda)S(\lambda)d\lambda}{\int_0^\infty S(\lambda)d\lambda}, \tag{1.12}$$

from which we can determine the filter-specific apparent magnitudes, as follows:

$$m_S = Z_S - 2.5 \log F_S. \tag{1.13}$$

For example, in the ultraviolet/optical/infrared windows, there are sets of standard filters, such as those shown in table 1.1. The choice of reference star – which gives the 'zero points' Z_S – defines the photometric system used. In each one, the reference standard has, by definition, zero magnitude through any passband. The most standard systems are:

1. VEGAMAG, where the reference star is Vega (αLyr);

2. AB, where the reference star has $F(\nu) = 3,631 \, \text{Jansky}$, constant at all frequencies;

Table 1.1 Some of the standard photometric passbands, showing the central wavelength (λ_0) and the full width at half maximum ($\Delta\lambda$) of the transmission function. For reference, we include the absolute magnitude of the Sun ($M_\odot$). Note that the *apparent* magnitude of the Sun from the Earth's position in the V band is $V_\odot = -26.74$.

Filter	U	B	V	R	I	J	H	K
$\lambda_0(\text{Å})$	3650	4400	5500	7000	9000	12500	16500	22000
$\Delta\lambda(\text{Å})$	680	980	890	2200	2400	3800	4800	7000
$M_\odot$	+5.61	+5.48	+4.83	+4.42	+4.08	+3.64	+3.32	+3.28

3. STMAG, for which the reference star has $F(\lambda) = 3.631 \times 10^{-9}\,\text{erg}\,\text{s}^{-1}\,\text{cm}^{-2}\,\text{Å}^{-1}$, constant at all wavelengths.

A colour index (or 'colour', in short) is defined as the ratio of flux densities of the same source, when observed through two different filters. In the logarithmic magnitude scale, the colour is therefore the difference of the magnitude in each of the filters; for instance,

$$U - V \equiv m_U - m_V = M_U - M_V = 2.5\left(\log \frac{L_V}{L_U} - \log \frac{L_V^0}{L_U^0}\right), \qquad (1.14)$$

where the 0 superindex refers to the reference star in a given photometric system. We use luminosities in this expression because the ratios of fluxes and luminosities should be the same. The colours of astrophysical sources provide useful information about their properties: the temperature of a star, the age and chemical composition of a stellar population, or the redshift of a distant galaxy. Colours contrasting different wavelength regions are sensitive to specific properties of the stellar populations or the gas and dust content. We can interpret a set of photometric colours as the low-resolution equivalent of a spectrum. Indeed, many studies are based on fluxes covering a large set of passbands that jointly provide a (multicolour) spectral energy distribution. The spectral resolution is $R \equiv \lambda/\Delta\lambda$, where λ is the (average) wavelength of the observation and $\Delta\lambda$ is the width of the passband (commonly quoted as the full width at half maximum of the spectral response function). From table 1.1 we find that typical broadband photometric studies have an effective resolution around $R \sim 5$. This is equivalent to smoothing the spectra shown in figure 1.2, reducing each spectrum to just five fluxes within the same wavelength interval. Although spectroscopy provides much more detailed information about galaxies, broadband photometry allows us to probe deeper, as we collect all the photons within the response of the filter into a single 'bin'. The standard approach in observational work is to survey and select interesting targets with photometry, following up a targeted set with spectroscopy, where the typical spectral resolution adopted when observing galaxies lies between R = 500 and 5,000. As a compromise, one can use medium-band or even narrow-band filters to increase the spectral resolution of photometric measurements (no higher than about 100 or 200) but allowing us to reach deeper than with spectroscopic observations. Note that the dynamical state of galaxies introduces a velocity dispersion of the stellar component $\sigma \gtrsim 50\,\mathrm{km\,s^{-1}}$. Therefore, a spectral resolution above $R = c/\sigma \sim 6,000$ is less useful for the analysis of, say, absorption line features originating in stellar atmospheres. Such a natural resolution limit imposed by the motion of stars in galaxies complicates, for instance, the study of detailed line strengths, which appear heavily blended in the spectra of galaxies with respect to data from individual stars.

Exercise 1.4

Consider two filters with a top-hat spectral response as shown in the figure. Compute the colour $S_1 - S_2$ for a star with a spectrum $F(\lambda) = k$, where k is a constant. Use the AB photometric system.

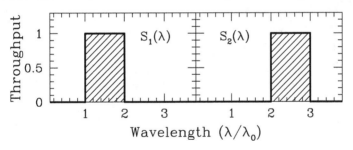

Notes

1 Baldwin, Phillips & Terlevich, 1981, PASP, 93, 5.
2 Caon, Capaccioli & D'Onofrio, 1993, MNRAS, 265, 1013.
3 Conselice, 2014, ARA&A, 52, 291.
4 Huertas-Company et al., 2015, ApJS, 221, 8.
5 Kennicutt, 1998, ARA&A, 36, 189.
6 Pasquali, 2015, *Astronomische Nachrichten*, 336, 505.
7 Gültekin et al., 2009, ApJ, 698, 198.

8   However, the cornerstone *Gaia* mission of the European Space Agency has
     shattered traditional measurements of parallax, with accuracies down to
     24 micro-arcseconds.
9   Bigiel et al., 2011, ApJ, 730, L13.
10 Kormendy & Kennicutt, 2004, ARA&A, 42, 603.
11 Vazdekis et al., 2012, MNRAS, 424, 157.
12 Thomas et al., 2005, ApJ, 621, 673.

2
The classical theory of gravitation

The Newtonian theory of gravitation is presented here, along with the most fundamental equations. We will begin with Kepler's problem, the simplest case of a gravitational system. The centrepiece of classical gravity as a field is Poisson's equation, which relates the mass density distribution with a scalar function (potential) from which the forces (and hence the dynamical evolution) can be followed. Variations of this field equation are found in many other areas of physics, including electrodynamics and general relativity. The concept of gravitational potential energy is presented. The importance of the potential/density relation is stressed in the chapter. Simple cases of potential theory are explored: point mass, leading to solar system dynamics; the isothermal sphere, often used to describe galaxies; the homogeneous sphere, which gives a good representation of the central region of our Galaxy; and more generic cases frequently used to describe the density profile of galaxies and dark matter halos. The projection of a volume density into an (observed) two-dimensional distribution is discussed, along with its inverse.

2.1 Gravitational force

The dominant force in the evolution of galaxies is gravitation. Over the lengthscales and densities typically found in galaxies, a Newtonian approach is well justified. Only in chapter 7 will we consider general relativity to follow the growth of density fluctuations in the framework of an expanding Universe. In its simplest form, the gravitational force on a mass m_1, exerted by a second mass, m_2, separated by a radius vector $\vec{r}_{12}$ (with origin at m_1) can be written

$$\vec{F}_{12} = -G\frac{m_1 m_2}{r_{12}^3}\vec{r}_{12}, \tag{2.1}$$

where G is the gravitational constant. This book deals with the role of this fundamental force in the formation and evolution of galaxies. We simplify the analysis by neglecting other mechanisms such as gas-related physics, energy loss due to electromagnetic interations or effects of two-body collisions. The latter will be justified in the next chapter.

Since forces from different sources can be added linearly, we can write the net gravitational force on mass m_1 exerted by a distribution of masses $\{m_i\}$ as

$$\vec{F}_1 = -Gm_1 \sum_{i\neq 1}\frac{m_i\vec{r}_{1i}}{r_{1i}^3} \rightarrow Gm_1 \iiint \frac{\vec{r}' - \vec{r}}{|\vec{r}' - \vec{r}|^3}\rho(\vec{r}')d^3r', \tag{2.2}$$

where the second expression takes the system to the limit of a continuous distribution, defined by a mass density $\rho(\vec{r})$. An important simplification can be invoked if the distribution of mass is spherically symmetric: $\rho(\vec{r}) = \rho(r)$. If the test particle, m_1, is located at a distance r from the centre of symmetry, we find that all the contributions to the net force from the mass outside r cancel out, leading to

$$\vec{F}_1 = -\frac{Gm_1}{r^2}\hat{e}_r \int_0^r \rho(s)4\pi s^2 ds. \tag{2.3}$$

Therefore, a test particle moving in a circular orbit with radius r around such a distribution of matter will have an orbital speed of

$$v_c^2(r) = \frac{GM(<r)}{r} = \frac{4\pi G}{r}\int_0^r \rho(s)s^2 ds. \tag{2.4}$$

Although stars in galaxies have a wide range of orbital shapes, this simple scaling relation is a powerful tool to give us an order of magnitude estimate, relating speed, mass and size. For instance, the Sun moves with an orbital speed $\sim$220 km s^{-1} around the Galactic centre, located $\sim$8 kpc away ($\sim$2.5 $\times$ 10^{20} m). Therefore, an order of magnitude guess for the Galactic mass is $M_{glx} \sim 9.1 \times 10^{10} M_\odot$ (where $M_\odot = 2 \times 10^{30}$ kg is the mass of the Sun), which is in the right ballpark. Hereafter, it may help to remember the value of the gravitational constant, G, in units more suitable for the analysis of galactic systems: $G = 4.3 \times 10^{-6}$ kpc(km/s)^2/$M_\odot$. It is also convenient to know that a speed of 1 km/s is approximately 1 pc/Myr.

2.2 The Kepler problem

The simplest case of a gravitational system is the Kepler problem, consisting of a star with mass m_s at position $\vec{r}_s$ and a planet with mass m_p at position $\vec{r}_p$. The forces acting on star and planet are, respectively:

$$
\left.
\begin{aligned}
m_s \ddot{\vec{r}}_s &= G\frac{m_s m_p}{|\vec{r}_s - \vec{r}_p|^3}(\vec{r}_p - \vec{r}_s) \\
m_p \ddot{\vec{r}}_p &= -G\frac{m_s m_p}{|\vec{r}_s - \vec{r}_p|^3}(\vec{r}_p - \vec{r}_s)
\end{aligned}
\right\}.
\tag{2.5}
$$

Hereafter, we follow the standard notation using a dot (two dots) over a quantity to denote its first (second) time derivative (e.g. $\ddot{x} = d^2 x/dt^2$). If we subtract one equation from the other, and define the relative vector separation as $\vec{r} \equiv \vec{r}_p - \vec{r}_s$, we simplify the two-body problem to a single gravitating system with mass $M = m_s + m_p$:

$$
\ddot{\vec{r}} = -\frac{GM}{r^3}\vec{r} \implies \frac{1}{2}\dot{r}^2 - \frac{GM}{r} = \text{constant} \equiv \mathcal{E},
\tag{2.6}
$$

where the second part is obtained by multiplying both sides by $\dot{\vec{r}}$ and then integrating with respect to time. This first integral represents energy conservation. A second conserved quantity can be obtained from the definition of angular momentum (per unit mass):

$$
\vec{\ell} \equiv \vec{r} \times \dot{\vec{r}} \implies \dot{\vec{\ell}} = \dot{\vec{r}} \times \dot{\vec{r}} + \vec{r} \times \ddot{\vec{r}} = 0,
\tag{2.7}
$$

which cancels for any central force (i.e., as long as the acceleration is aligned with the radius). Therefore $\vec{\ell}$ is a constant vector, and there is no loss of generality if we assume the whole trajectory to be confined on the XY plane, so that the total angular momentum is: $\vec{\ell} = \ell \hat{e}_z$, with ℓ a constant. Noting that ℓ is the area of the parallelogram spanned by $\vec{r}$ and $\dot{\vec{r}}$, we arrive at Kepler's second law: *The radius vector joining the star with the planet sweeps out equal areas during equal intervals of time.* The area swept per unit time is $dA/dt = \ell/2$. Using cylindrical coordinates (R, θ, z) allows us to write the energy (per unit mass) as

$$
\mathcal{E} = \frac{\dot{R}^2}{2} + \Phi_{\text{eff}}(R), \quad \text{where} \quad \Phi_{\text{eff}}(R) = \frac{\ell^2}{2R^2} - \frac{GM}{R}.
\tag{2.8}
$$

Note this is an equivalent one-dimensional problem involving the radial distance from star to planet.

Exercise 2.1

Show that the orbit in a Keplerian potential can be obtained by integrating equation 2.8 with respect to the azimuthal angle θ, and defining $u \equiv 1/R$, so that $\dot{R} = -\ell du/d\theta$, leading to

$$R = \frac{\alpha}{1 + e\cos\theta},$$

which corresponds to a conic section with eccentricity e:

$$e = \sqrt{1 + \frac{2\mathcal{E}\ell^2}{G^2 M^2}},$$

and $\alpha = \ell^2/GM$ is the semilatus rectum, directly related to the size of the orbit, $\alpha = b^2/a$, where a and b are the semi-major and semi-minor axes of the conic section, respectively.

Exercise 2.1 confirms Kepler's first law – *The orbit of the planet is a conic section with the star at one of the two foci* – adding to it how the orbital parameters are related to the mass of the star and the energy and angular momentum of the orbit. Finally, for a closed orbit, we can derive its period either by integrating the energy equation (2.8) with respect to time, or by using Kepler's second law, noting that the area of an ellipse is πab (see exercise 2.1), and that therefore the period is: $\tau = 2\pi ab/\ell = 2\pi a\sqrt{\alpha a}/\ell$, where substituting α with respect to mass and angular momentum gives

$$GM = \omega^2 a^3, \tag{2.9}$$

where $\omega = 2\pi/\tau$ is the angular frequency. Equation 2.9 is the mathematical expression of Kepler's third law:[1] *The square of the orbital period of the planet is proportional to the cube of the semi-major axis of its orbit.* Moreover, we directly relate this period-orbital size relation to the mass of the system.

We can use the effective potential, defined in equation 2.8, to illustrate the expected shape of the orbits. The Kepler problem produces orbits that are conic sections. The type of orbit depends on the initial conditions – that is, the position and velocity given to the planet 'in the beginning'. We can use an energy-based argument to determine these orbits. Note that equation 2.8 corresponds to a one-dimensional case, namely, following the radial distance from the planet to the star. In addition, the

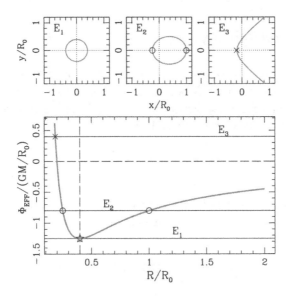

Figure 2.1 Kepler's orbits and the gravitational potential. The bottom panel shows the effective potential for a choice of angular momentum ($\ell^2 = 2GMR_0/5$), and three possible types of orbits defined by the total energy: E_1, E_2 and E_3. The orbit for each case is shown in the top three panels.

conserved angular momentum gives us the azimuthal component, via: $\dot{\theta} = v_\perp/R = \ell/R^2$. Figure 2.1 shows the effective potential for a specific choice of angular momentum, and three possible cases for the total energy: E_1, E_2, E_3. Noting that the kinetic energy term is positive definite, we find that at the minimum (E_1) there is only one possible radial distance, so that the planet describes a circular orbit with constant speed (and zero eccentricity, i.e., $E_1 = -G^2M^2/2\ell^2$). In the region $E_1 < E < 0$, the planet is also bound, this time between an aphelion and a perihelion, tracing an elliptical orbit. If $E > 0$, the orbit corresponds to a hyperbola, and at the limiting case $E = 0$, the planet traces a parabola (with eccentricity $e = 1$).

Exercise 2.2

Find the equivalent of Kepler's three laws of motion if the gravitational force were to behave like a simple harmonic oscillator; i.e., substitute equation 2.1 with

$$\vec{F}_{12} = -Gm_1m_2\vec{r}_{12}.$$

2.3 Potential theory

The gravitational force is a vector field: at each point in space, the acceleration on a test particle is uniquely described by a vector. The gravitational potential is a scalar function that provides the same amount of information as the vector field, and it is defined as

$$\Phi(\vec{r}) \equiv -G \iiint \frac{\rho(\vec{r}')}{|\vec{r}' - \vec{r}|} d^3 r'. \tag{2.10}$$

One can show that at each point in space, the gradient of this scalar function produces a vector field equivalent to the gravitational force. By using the expression

$$\vec{\nabla}_{\vec{r}} \left(\frac{1}{|\vec{r}' - \vec{r}|} \right) = \frac{\vec{r}' - \vec{r}}{|\vec{r}' - \vec{r}|^3}, \tag{2.11}$$

we can write the gravitational force on a test particle with mass m as

$$\frac{\vec{F}(\vec{r})}{m} = -\vec{\nabla}_{\vec{r}} \Phi(\vec{r}). \tag{2.12}$$

Moreover, we can derive a field equation that relates the gravitational potential to the mass density. Taking the divergence of equation 2.12:

$$\vec{\nabla}_{\vec{r}} \cdot \left(\frac{\vec{F}}{m} \right) = G \iiint \vec{\nabla}_{\vec{r}} \cdot \left(\frac{\vec{r}' - \vec{r}}{|\vec{r}' - \vec{r}|^3} \right) \rho(\vec{r}') d^3 r', \tag{2.13}$$

and using the following identity:

$$\vec{\nabla}_{\vec{r}} \cdot \left(\frac{\vec{r}' - \vec{r}}{|\vec{r}' - \vec{r}|^3} \right) = -\frac{3}{|\vec{r}' - \vec{r}|^3} + \frac{3(\vec{r}' - \vec{r}) \cdot (\vec{r}' - \vec{r})}{|\vec{r}' - \vec{r}|^5}, \tag{2.14}$$

we note that this expression vanishes when $|\vec{r}' - \vec{r}| \neq 0$, so that any contribution to the integral must come from an infinitesimal volume around $\vec{r}' = \vec{r}$. Therefore, we restrict the volume of integration to a small sphere at $\vec{r}' = \vec{r}$ with radius $h \to 0$. Within the infinitesimal volume of integration, we can swap $\vec{r}$ and $\vec{r}'$ in the gradient operator, allowing us to apply the divergence theorem, and changing the volume integral into a surface integral:

$$\vec{\nabla}_{\vec{r}} \cdot \left(\frac{\vec{F}}{m} \right) = -G\rho(\vec{r}) \oiint_{|\vec{r}' - \vec{r}| = h} \frac{(\vec{r}' - \vec{r}) \cdot d^2 \vec{S}}{|\vec{r}' - \vec{r}|^3}. \tag{2.15}$$

On the sphere, the surface element is related to the solid angle element by $d^2\vec{S} = (\vec{r}' - \vec{r})hd^2\Omega$, and therefore

$$\vec{\nabla}_{\vec{r}} \cdot \left(\frac{\vec{F}}{m}\right) = -G\rho(\vec{r}) \oiint d^2\Omega = -4\pi G\rho(\vec{r}). \qquad (2.16)$$

If we write the divergence of the force with respect to the scalar potential, we arrive at the gravitational field equation:

$$\nabla^2 \Phi(\vec{r}) = 4\pi G\rho(\vec{r}), \qquad (2.17)$$

also called 'Poisson's equation'. Note this equation provides a direct link between the source of the field (the density distribution) and the gravitational forces, described by $\Phi(\vec{r})$. This is the Newtonian equivalent of Einstein's general relativistic equation – $G_{\mu\nu} = 8\pi G T_{\mu\nu}$ – linking the field (through the geometric tensor, $G_{\mu\nu}$) to the mass-energy distibution (through the stress-energy tensor, $T_{\mu\nu}$).

Newton's theorems

Two theorems proved by Newton for spherically symmetric mass distributions provide a simple way to solve for the gravitational potential. The first theorem states that the gravitational force on a test mass inside a spherical shell with constant surface density is zero. The second theorem proves that the force on a test mass outside this shell is the same as if the whole shell mass were reduced to a point mass at the centre.

There are several ways to solve the theorems. Let us consider a test point mass P, inside the shell. Figure 2.2 shows how to solve it via solid angles. We compute the gravitational force exerted by the masses within two opposite and infinitesimal regions of the shell, which subtend the same solid angle $d\Omega$. When Σ is defined as the (constant) surface mass density of the shell, the mass from each of these two opposed regions is $dm_1 = \Sigma r_1^2 d\Omega / \cos\theta$ and $dm_2 = \Sigma r_2^2 d\Omega / \cos\theta$. Note the angles are the same, as the triangle formed by the centres of these two mass elements and the origin is isosceles. Hence, the forces per unit mass are

$$\left.\begin{aligned}
\frac{F_1}{m} &= \frac{dm_1}{r_1^2} = \frac{\Sigma d\Omega}{\cos\theta} \\
\frac{F_2}{m} &= \frac{dm_2}{r_2^2} = \frac{\Sigma d\Omega}{\cos\theta}
\end{aligned}\right\}. \qquad (2.18)$$

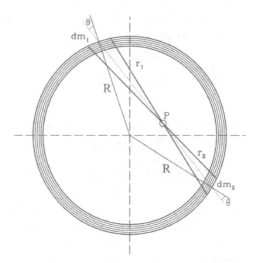

Figure 2.2 Solution to Newton's theorem, using a simple geometric argument. Note that at an arbitrary point, represented by P, the solid angles subtended by dm_1 and dm_2 are equal.

Since they have the same magnitude and opposite direction, they will cancel out. Integrating throughout all solid angles, we obtain a net zero force.

We can solve the second theorem by invoking Gauss's law, making the assumption that, because of the spherical symmetry of the mass distribution, the potential (or the magnitude of the force) can depend only on the radial distance r. Choosing a spherical surface S, at distance r from the origin, enclosing a volume, $\mathcal{V}$, that includes the shell, we find

$$\oiint_{\partial S(r_0)} \vec{F}(r) \cdot d\vec{\Sigma} = \iiint_{\mathcal{V}(r_0)} \vec{\nabla} \cdot \vec{F}(r) d\mathcal{V} = -\iiint_{\mathcal{V}(r_0)} \nabla^2 \Phi(r) d\mathcal{V} =$$

$$= -4\pi G \iiint_{\mathcal{V}(r_0)} \rho(r) d\mathcal{V} = -4\pi GM, \tag{2.19}$$

where we also apply Poisson's equation (2.17). Since the force at S is the same in magnitude, we obtain $F(r) = GM/r^2$, as expected if all the mass of the shell were located at the origin.

Show that Newton's first theorem can be solved by integrating throughout the spherical shell. An alternative method involves showing that the gravitational potential inside any point inside the shell is constant.

2.4 Gravitational potential energy

We have seen in section 2.3 that the gravitational scalar potential gives a complete description of the force vector field, via $\vec{F} = -\vec{\nabla}\Phi$. Such a result implies that the work involved in moving a test particle along a closed orbit ($\mathcal{C}$) should be zero:

$$\text{Work} = \oint_{\mathcal{C}} \vec{F} \cdot d\vec{r} = \iint_S (\vec{\nabla} \times \vec{F}) \cdot d\vec{S} = -\iint_S (\vec{\nabla} \times \vec{\nabla}\Phi) \cdot d\vec{S} = 0, \quad (2.20)$$

where we have applied Stokes's theorem to the line integral and used the fact that the curl of a gradient must vanish. This property characterizes conservative forces and allows us to derive the gravitational potential energy. Let us assume we have a point mass m at the centre of coordinates, and we bring another mass m from infinity, assumed to be initially at rest. The net change in total energy will be the work done to bring that particle close to the mass at $r = 0$, namely:

$$\text{Work} = \int \vec{F} \cdot d\vec{r} = \int_{\infty}^{r} \frac{Gm^2}{r^2} dr = -\frac{Gm^2}{r} = \Delta E, \quad (2.21)$$

where we assume the particle's final position at radial distance r from the origin. This net change of energy can be associated with an increase in the gravitational potential energy: $\Delta E = \text{Work}$. If we do the same with an ensemble of masses, we have to add the pairwise contributions to the energy change, making sure we do not count twice the same pair, namely:

$$W = -\frac{Gm^2}{2} \sum_{i \neq j} \frac{1}{r_{ij}} = \frac{1}{2} \sum_i m\Phi(\vec{r}_i), \quad (2.22)$$

where we follow the standard notation using W as the gravitational potential energy, and define r_{ij} as the separation between the ith and the *j*th mass. The second part of the equation takes advantage of the definition of the gravitational potential seen by the *i*th particle. This expression can be extended to a continuous distribution of mass by replacing the sum with a volume integral, using densities instead of masses:

$$W = \frac{1}{2} \iiint \rho(\vec{r})\Phi(\vec{r})d^3r. \tag{2.23}$$

Based on dimensional grounds, it is often useful to write this expression with respect to the total mass of the system (*M*) and some estimate of its size (*R*):

$$W = -\gamma \frac{GM^2}{R}, \tag{2.24}$$

where the so-called fudge factor (γ) is just a numerical value, typically not very different from 1.

Exercise 2.4

Show that the gravitational potential energy of a sphere with mass *M*, radius *R*, and constant density is

$$W = -\frac{3}{5}\frac{GM^2}{R}.$$

Assuming that the solar luminosity $L_\odot \sim 4 \times 10^{33}$ erg s^{-1} is caused by a slow contraction of the star releasing gravitational potential energy into heat, derive the expected lifetime of the Sun, and contrast your result with its current age $\sim$4.5 Gyr.

2.5 Potential/density pairs: A few fundamental cases

Poisson's equation (equation 2.17) implies that for any mass distribution, we can define a gravitational potential from which the force on a test mass can be obtained. It is useful to define below a set of density-potential pairs that will be very relevant to galaxy dynamics. We assume here spherical symmetry. Moreover, it is also practical to derive the radial dependence of the orbital speed of a test mass in a circular orbit. One of the key observables in galaxies and stellar clusters is the velocity field, from which we can derive a large amount of information about the underlying potential.

The circular speed gives us a good order of magnitude estimate, and in systems where the kinetic energy is dominated by rotation – such as disc galaxies – it allows us to probe the mass distribution.

Point mass: This is the Kepler problem described above, for which the density-potential pair is

$$\rho(r) = M\delta(r) \longleftrightarrow \Phi(r) = -\frac{GM}{r}, \tag{2.25}$$

from which we derive an orbital speed $v^2(r) = GM/r$, which corresponds to the characteristic orbital speeds of planets in the solar system (where $M \approx M_\odot$). This is also the typical velocity profile of matter (gas and stars) moving close to a supermassive black hole, as found in the centres of galaxies. A Keplerian profile is a telltale signature of a highly concentrated distribution of mass.

Homogeneous sphere: The density/potential pair for a sphere with radius a and constant density ρ_0 is

$$\rho(r) = \rho_0 H(a - r) \longleftrightarrow \Phi(r) = \begin{cases} -2\pi G\rho_0(a^2 - \frac{1}{3}r^2) & r \leq a \\[2mm] -\dfrac{4\pi G\rho_0 a^3}{3r} & r > a. \end{cases} \tag{2.26}$$

The orbital speed inside the mass distribution rises linearly with radius $v = r\sqrt{4\pi G\rho_0/3}$, switching over to a Keplerian profile at $r > a$. Inside the sphere, the potential is equivalent to a three-dimensional harmonic oscillator, with the same spring constant along the three directions, and an associated period: $\tau = \sqrt{3\pi/G\rho_0}$. Note that the dynamical timescale of a gravitational system is $t_{\text{dyn}} \propto \sqrt{1/G\rho}$.

Isothermal sphere: An isothermal sphere is a density distribution often encountered in gravitating systems. We will see in chapter 3 that it corresponds to the galaxy dynamics equivalent of a system in thermodynamic equilibrium (although there are substantial differences, as we will see in that chapter). The density/potential pair is

$$\rho(r) = \rho_0 \left(\frac{r}{a}\right)^{-2} \longleftrightarrow \Phi(r) = 4\pi G\rho_0 a^2 \ln(r/a), \tag{2.27}$$

and the orbital speed is $v_c^2 = 4\pi G\rho_0 a^2$, i.e., constant. Both the potential and the total mass diverge at large radii. A more realistic version of this case involves a truncated profile, where the density vanishes at $r > a$, and the total mass is therefore $M = 4\pi\rho_0 a^3$, with the potential outside the sphere being Keplerian: $\Phi(r > a) = -GM/r$.

Exercise 2.5

A slight variation of the isothermal sphere is the Jaffe model, where the gravitational potential is given by

$$\Phi(r) = 4\pi G\rho_0 a^2 \ln\left(\frac{x}{1+x}\right),$$

and $x \equiv r/a$. Show that the density of this model is

$$\rho(r) = \rho_0 \frac{1}{x^2(1+x)^2},$$

and the total mass is $M = 4\pi\rho_0 a^3$.

NFW profile: The first computer simulations of galaxy formation explored the growth of structure under a gravitational potential in an expanding Universe. As we will see in chapter 7, small density fluctuations grow under gravity, creating stable (virialized) structures called 'halos'. The numerical simulations appeared to settle into a universal density profile, termed 'NFW' after the authors of the seminal work.[2] An NFW distribution is defined by a total (virial) radius (r_{VIR}), a scale radius (a), and a central density ρ_0, giving the following density/potential pair:

$$\rho(x) = \frac{\rho_0}{x(1+x)^2} \longleftrightarrow \Phi(x) = -\frac{4\pi G\rho_0 a^2}{x} \ln(1+x), \qquad (2.28)$$

where the dimensionless radial coordinate: $x \equiv r/a$. The circular speed in this potential is

$$v_c(x) = \frac{4\pi G\rho_0 a^2}{x}\left[\ln(1+x) - \frac{x}{1+x}\right]. \qquad (2.29)$$

The logarithmic slope $\gamma \equiv d(\ln \rho)/d\ln r$ of the NFW profile goes from -1 at $r \to 0$ to -3 at $r \to \infty$. At the scale length $r = a$, the slope reaches the value of the isothermal slope, $\gamma = -2$. Note that since the density scales like $\rho \propto r^{-3}$ at large radii, the cumulative mass diverges logarithmically.

Therefore, we also need to truncate the distribution. The outer radius is commonly defined as the virial radius, i.e., the region within which all particles are bound in virial equilibrium (see section 3.9). The dimensionless concentration parameter (c) is defined as the ratio between the virial radius and the scale radius. This is one of the most studied density distributions, given its relevance in the study of dark matter halos. An alternative definition includes free parameters for the behaviour at small and large radii in the so-called generalized NFW profile:

$$\rho(x) = \frac{\rho_0}{x^\alpha (1+x)^\beta}, \tag{2.30}$$

where $(\alpha, \beta) = (1, 2)$ reduces to the NFW profile, $(2, 0)$ gives the isothermal sphere, and $(2, 2)$ is the Jaffe model presented in exercise 2.5.

Hernquist profile: A similar density distribution was proposed to explain the mass distribution in elliptical galaxies.[3] In this case, we have

$$\rho(x) = \frac{\rho_0}{x(1+x)^3} \longleftrightarrow \Phi(x) = -\frac{2\pi G\rho_0 a^2}{1+x}. \tag{2.31}$$

The two-dimensional projection of this density profile follows quite closely the standard de Vaucouleurs ($R^{1/4}$) surface brightness profile, assuming a constant M/L ratio. The logarithmic slope changes from -1 to -4; therefore the total mass of the distribution is finite ($M = 2\pi\rho_0 a^3$), as the density decreases faster than the volume at large radii. The circular speed is

$$v(x) = \sqrt{\frac{GM}{a}} \frac{x^{1/2}}{1+x}. \tag{2.32}$$

Plummer sphere: This potential is especially important in numerical simulations of dynamical systems, where the parameter λ (softening length) eliminates the problem of numerical divergence when computing the potential for small values of r. In this case, λ defines the resolution limit in the calculation of gravitational forces. The density/potential pair is

$$\rho(r) = \left(\frac{3M}{4\pi\lambda^3}\right)\left(1 + \frac{r^2}{\lambda^2}\right)^{-5/2} \longleftrightarrow \Phi(r) = -\frac{GM}{\sqrt{r^2 + \lambda^2}} \tag{2.33}$$

Note that as $\lambda \to 0$, the Plummer sphere resembles a Keplerian profile. At small radii ($r \ll \lambda$), the model mimics the inside of a sphere with constant density.

King profile: This model is often adopted to describe the mass distribution of stellar clusters, elliptical galaxies and even galaxy clusters. It gives a good fit to a numerical method that assumes a Maxwellian velocity distribution along with a simple prescription to take into account the ejection of stars moving quickly enough to become unbound (see section 8.4):

$$\rho(r) = \rho_0 \left[1 + \left(\frac{r}{r_c} \right)^2 \right]^{-3/2} \longleftrightarrow$$

$$\Phi(r) = \Phi_0 + 4\pi G \rho_0 r_c^2 \left[1 - \frac{\sinh^{-1}(r/r_c)}{(r/r_c)} \right]. \qquad (2.34)$$

Yukawa potential: The Yukawa potential is defined as the Keplerian case with an exponential cutoff over a lengthscale r_0, namely:

$$\rho(r) = -\frac{GM}{r_0^2 r} e^{-r/r_0} \longleftrightarrow \Phi(r) = -\frac{GM}{r} e^{-r/r_0}. \qquad (2.35)$$

The Poisson equation for this case gives a negative density! This merely states that such a mass distribution that results in exponentially decaying forces at large distances cannot be accommodated in gravitation; i.e., the force of gravity cannot be screened. We will see in chapter 3 that this is one of the main reasons why we can consider the motion of a star within a large ensemble as produced by the gravitational potential produced by the general distribution of stars. The Yukawa potential gives a valid description of weak interactions (as in those felt by neutrinos[4]).

Notice that in many of the potentials presented here, the density profile diverges at small radii. This does not cause a divergence in the amount of mass in the central region, as long as the rate of increase of the density is slower than the decrease of volume as $r \to 0$. However, divergent central densities (cusps) may be erased by other mechanisms, such as the infall of smaller structures (mergers), creating a flatter distribution: a core. The core-cusp dilemma in galaxy halos is one of the open problems today.

Exercise 2.6

Find the circular velocity profile of the last three cases (Plummer, King, Yukawa).

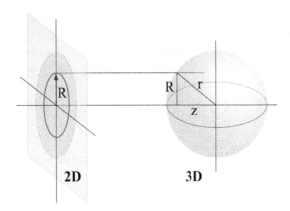

Figure 2.3 Projection of a spherical distribution onto a two-dimensional (observer) plane.

2.6 Two-dimensional projection

A comparison of a density distribution with an observed galaxy or cluster involves the projection of the three-dimensional volume mass density into a two-dimensional surface mass density (or a deprojection if going from the observational to the theoretical data). To transform mass into light (needed when contrasting models with observational data), we need to include the mass to light ratio, $\Upsilon \equiv M/L$. One typically assumes a simple scaling of Υ with respect to the galactocentric radius. Let us consider here the simple case of a spherical galaxy with constant density $\rho(r) = \rho_0$ inside $r < r_0$, vanishing outside. We need to integrate along the line of sight (figure 2.3) to write the 2D surface brightness, $\Sigma(R)$, as a function of the 3D density, $\rho(r)$:

$$\Sigma(R) = \int_{-\infty}^{+\infty} \Upsilon^{-1}(r)\rho(r)dz = \int_{R}^{r_0} \frac{2r\rho(r)}{\Upsilon(r)\sqrt{r^2 - R^2}}dr = \frac{2\rho_0}{\Upsilon_0}\sqrt{r_0^2 - R^2}.$$
(2.36)

The last step makes the approximation of a constant mass-to-light ratio throughout the galaxy. The inverse process, i.e., the deprojection from a 2D surface brightness profile to a 3D mass density distribution, can also be obtained by considering that equation 2.36 describes an Abel transform that can be inverted as follows (assuming a constant Υ):

$$\rho(r) = -\frac{\Upsilon_0}{\pi}\int_{r}^{\infty} \frac{d\Sigma}{dR}\frac{dR}{\sqrt{R^2 - r^2}}.$$
(2.37)

Of course, in practice, this expression may not be practical if the data are too noisy or the assumption of some level of symmetry is unrealistic. One useful pair of 2D/3D distributions corresponds to the modified Hubble law:

$$\frac{\Sigma(R)}{\Sigma_0} = \left[1+\left(\frac{R}{r_0}\right)^2\right]^{-1} \longleftrightarrow \frac{\rho(r)}{\rho_0} = \left[1+\left(\frac{r}{r_0}\right)^2\right]^{-3/2}, \qquad (2.38)$$

where $\Sigma_0 = 2\rho_0 r_0/\Upsilon_0$. Note that at large radii, the surface brightness decreases as $\propto R^{-2}$ and the mass density scales like $\propto r^{-3}$. In either case, the integrated quantity diverges logarithmically.

Exercise 2.7

Consider a galaxy described by a spherically symmetric Hernquist profile (equation 2.31), and assume a constant mass to light ratio, Υ. Show that the two-dimensional projected surface brightness of this galaxy is

$$\Sigma(y) = \frac{M}{\Upsilon 2\pi a^2} \frac{(y^2+2)F(y)-3}{(1-y^2)^2},$$

where $y \equiv R/a$, and R is the projected radial coordinate, and

$$F(y) = \begin{cases} \dfrac{1}{\sqrt{1-y^2}}\,\mathrm{sech}^{-1}y, & \text{for } 0 \leq y \leq 1 \\[2ex] \dfrac{1}{\sqrt{y^2-1}}\,\mathrm{sec}^{-1}y, & \text{for } 1 \leq y < \infty. \end{cases}$$

Notes

1 Aptly called Kepler's '1-2-3' law in Misner, Thorne & Wheeler, 1973, *Gravitation*, W. H. Freeman.
2 Navarro, Frenk & White, 1997 ApJ, 490, 493.
3 Hernquist, 1990, ApJ, 356, 359.
4 E.g., Bettini, 2008, *Introduction to elementary particle physics*, Cambridge.

3
A statistical treatment of stellar systems

A many-body gravitational problem – involving millions or even billions of stars – cannot be solved in the same way as the standard two-body problem. A statistical treatment is required, following an analogy between thermodynamics and statistical physics. The concept of phase space and the distribution function is fundamental to a statistical description of a dynamical system. This treatment is valid for stellar systems because of the nature of gravity and the characteristic densities in these systems, which lead to a long relaxation time, a concept that will be explored in some detail here. The main equation that governs the evolution of the distribution function in a system made up of stars (or dark matter particles) is the collisionless Boltzmann equation (CBE). Simplifications to the CBE arise from symmetries of the system (similarly to Noether's theorem in theoretical physics), leading to the concept of isolating integrals and Jeans theorem and allowing us to describe the distribution function in terms of a reduced set of variables (the isolating integrals themselves). Simple cases are discussed with spherical and axial symmetry. A more pragmatic approach to the CBE involves taking moments of velocity, leading to Jeans equations, a fundamental workhorse in stellar dynamics. A brief glimpse of the consequences of perturbatively removing the collisionless behaviour of stars is presented, with dynamical friction being one of the most representative effects.

3.1 Phase space

Since the evolution of a dynamical system is governed by a second order differential equation of the position with respect to time (i.e., $\vec{F} = m\ddot{\vec{r}}$), there are two boundary conditions, say at some reference time (t_0): on the position $\vec{r}(t_0)$, and on the velocity $\vec{v}(t_0) = \dot{\vec{r}}(t_0)$. Therefore, the evolution can be traced in phase space, comprised of position and velocity (or linear momentum). Phase space is therefore six dimensional. Let us consider an example of one-dimensional motion subject to a simple harmonic potential: $\Phi(x) = \frac{1}{2}kx^2$. The solution to the equation of motion of a particle of mass m in this potential is

$$\left.\begin{array}{l} x(t) = A\sin(\omega t + \phi_0) \\ \dot{x}(t) = A\omega\cos(\omega t + \phi_0) \end{array}\right\}, \tag{3.1}$$

where the angular frequency is $\omega^2 = k/m$; A is the amplitude of the oscillations; and ϕ_0 is a phase term that depends on the initial position and velocity of the mass. The trajectory in the corresponding two-dimensional phase space is an ellipse. Each ellipse has a unique value of the (conserved) energy, $E = mA^2\omega^2$, and the trajectories corresponding to different values of the energy do not cross, creating a nested set of ellipses that cover all of phase space. This is an important concept that will be exploited when we discuss Jeans theorem below.

Exercise 3.1

Find the trajectories in phase space of closed orbits in the Kepler problem (see section 2.2). Simplify the 4-dimensional parameter space – we can neglect motion along the z-axis – by plotting subspaces $(R, \dot{R})$ and $(\theta, \dot{\theta})$.

3.2 The distribution function

Now we change from a single star moving under a potential, to a many-body system comprised of multiple stars. One can study the evolution of all the stars in phase space by creating a complex set of trajectories. We define the distribution function $-f(\vec{r}, \vec{v}; t)$ – as the number of stars located within some differential volume element in phase space ($d\tau = d^3r\,d^3v$), at some time, t (hereafter, we use DF to refer to the distribution function). After normalization, the distribution function can be considered a probability distribution in phase space. For instance, we define the spatial number density of stars as

$$v(\vec{r};t) \equiv \iiint f(\vec{r},\vec{v};t)d^3v. \tag{3.2}$$

Moreover, we can define observables such as the average velocity as

$$\langle \vec{v}(\vec{r};t) \rangle \equiv \frac{1}{v(\vec{r};t)} \iiint f(\vec{r},\vec{v};t)\vec{v}d^3v, \tag{3.3}$$

or the velocity dispersion (via the variance) as

$$\sigma_v^2(\vec{r};t) \equiv \langle v^2(\vec{r};t) \rangle - \langle \vec{v}(\vec{r};t) \rangle^2$$

$$= \frac{1}{v(\vec{r};t)} \iiint f(\vec{r},\vec{v};t)v^2 d^3v - \langle \vec{v}(\vec{r};t) \rangle \cdot \langle \vec{v}(\vec{r};t) \rangle. \tag{3.4}$$

These will be useful quantities that can be compared with observations. For instance, the projection of the average velocity can be extracted from spectroscopic observations of galaxies, where the spectral lines appear differentially offset depending on their bulk motion, due to the combination of Doppler-shifted lines for every star in the system. Likewise, the velocity dispersion can be inferred from the widths of the spectral lines. Most of the time, we assume a steady state, where the distribution functions and all the averaged quantities do not have any explicit time dependence.

3.3 Relaxation time

There is a direct analogy between the motion of stars in a galaxy and the motion of particles in a fluid, say, e.g., molecules of nitrogen in a room. Although it is possible to envisage the trajectory of a molecule in phase space, the number of collisions (understood as interactions with other molecules that affect its dynamical state) is so numerous that after a very short timescale, these interactions have completely altered its initial dynamical state. This process, called 'relaxation', affects all molecules and erases all information about their initial dynamical state. This is the main reason why it is relatively simple to deal with a complex system made up of a large number of particles as a macroscopic (thermodynamic) ensemble. In a relaxed system, energy equipartition is achieved, and, in terms of the central limit theorem, the distribution function of energies can be described as a Gaussian distribution, with a simple parameter that controls its width:

$$f(\vec{r},\vec{v}) \propto e^{-\frac{mv^2}{2kT}}. \tag{3.5}$$

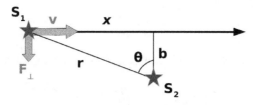

Figure 3.1 Computing the relaxation time: simplified (rectilinear) trajectory of a star-star encounter.

This is the Maxwell-Boltzmann distribution. The single parameter that uniquely defines this system, temperature, enables us to wholly describe its properties. Is it possible to extend this treatment to the distribution of stars moving in a galaxy? To answer that question, we need to consider the relaxation timescale in typical gravitating systems.

We can address this problem by studying the trajectory of a star (S_1) moving in the galaxy. We begin with a single encounter with another star, S_2 (figure. 3.1). Both stars have mass m. The point of closest approach, b, is termed the 'impact parameter'. We assume here that the deflection angle is small, so that the trajectory is approximately a straight line. The gravitational force of S_2 on S_1 is decomposed into longitudinal and tangential components. There is a net change in the tangential component of the velocity from the force:

$$F_\perp = F\cos\theta = \frac{Gm^2}{b^2+x^2}\frac{b}{r} = \frac{Gm^2}{b^2}\left[1+\left(\frac{vt}{b}\right)^2\right]^{-3/2}. \qquad (3.6)$$

This net force will result in a change of the tangential component of velocity:

$$\delta v_\perp = \frac{Gm}{b^2}\int_{-\infty}^{+\infty}\left[1+\left(\frac{vt}{b}\right)^2\right]^{-3/2}dt = \frac{2Gm}{bv}. \qquad (3.7)$$

This change in velocity can be interpreted as the force at the point of closest distance ($r=b$) multiplied by the 'duration' of the passage (b/v), i.e., an impulse. The symmetry of the process results in a cancellation of the longitudinal component during the passage ($\delta v_\parallel = 0$). Now we assume there are a number of similar passages as S_1 moves throughout the galaxy. Let us consider the galaxy as a flat system, with N stars within a radius R. After a crossing time – defined simply as a typical timescale for a star to traverse the length of the galaxy – the number of encounters with impact parameter b within an infinitesimal interval db – can be written

$$\delta n = \frac{N}{\pi R^2}2\pi b\, db. \qquad (3.8)$$

The collisions are random events, so that, on average, there is no net change in the velocity ($\langle \delta v_\perp \rangle = 0$). However, the interactions will impart a level of fluctuations, as expected from a random walk process, with a root mean square given by

$$\delta v_\perp^2 = \left(\frac{2Gm}{bv}\right)^2 \frac{2N}{R^2} b\,db. \tag{3.9}$$

The net effect will be obtained by integrating across all impact parameters. However, as is always the case in these scattering processes, the contribution diverges as $b \to 0$. We simply substitute the lower limit by some $b_{\min}$, at which our simple approximation of a linear trajectory breaks down. $b_{\min}$ can thus be defined as the value of b for which the change in velocity is as large as the actual speed of the incoming star: $\delta v_\perp \sim v \Longrightarrow b_{\min} = Gm/v^2$. We also change the upper limit to the extent of the galaxy and get

$$\Delta v_\perp^2 = \int_{b_{\min}}^{R} \delta v_\perp^2 \, db = 8N \left(\frac{Gm}{Rv}\right)^2 \ln\left(\frac{R}{b_{\min}}\right). \tag{3.10}$$

Note that the effect of the integration limits appear as a (weakly varying) logarithmic trend, usually called the 'Coulomb logarithm' ($\ln \Lambda$). Just to get a feeling for the effect of this term, for typical galaxy scales ($R \sim 10\,\text{kpc}$; $m = M_\odot$; $v \sim 200\,\text{km/s}$) we get $b_{\min} \sim 0.02\,\text{AU}$. Interactions within this value of b are very unlikely unless the system has density $\sim 1/b_{\min}^3$, which is too high in all stellar systems. Therefore the approximation is well justified.

Finally, in order to assess the onset of relaxation, we would need to have this level of fluctuation in velocity comparable with a typical velocity in the galaxy, roughly $v^2 = GM/R = GNm/R$, where M is the mass of the galaxy. Therefore, when relaxation is achieved, after n_{rel} 'crossings' of the galaxy by the star,

$$n_{\text{rel}} \left(\frac{\Delta v_\perp}{v}\right)^2 = n_{\text{rel}} \frac{8 \ln \Lambda}{N} \sim 1. \tag{3.11}$$

The argument of the Coulomb logarithm can be simplified as follows:

$$\Lambda = \frac{R}{b_{\min}} = \frac{R}{Gm} v^2 \sim N, \tag{3.12}$$

leading to

$$n_{\text{rel}} \sim \frac{N}{8 \ln N}. \tag{3.13}$$

Table 3.1 Relaxation times of typical stellar systems.

System	N	R pc	v km/s	t_{cross} yr	t_{rel} yr	Age/t_{rel}
Open cluster	100	2	0.5	4×10^6	10^7	≤ 1
Globular cluster	10^5	4	10	4×10^5	4×10^8	≤ 10
Massive galaxy	10^{11}	10^4	300	3×10^7	10^{17}	10^{-7}
Dwarf galaxy	10^9	10^3	50	2×10^7	10^{14}	10^{-4}
Galaxy cluster	10^3	10^6	10^3	10^9	2×10^{10}	0.1

This expression can be translated to a timescale by use of the crossing time:

$$t_{rel} = n_{rel}\, t_{cross} = \frac{N}{8 \ln N} \frac{R}{v}. \tag{3.14}$$

Table 3.1 shows the relaxation times in several stellar systems. In most cases, the relaxation time is comparable or even greater than the current age of the Universe. Only in the densest regions (e.g., the cores of globular clusters) could one expect the effects from relaxation to be significant. In fact, the inherent instability of gravitational forces (unscreened, as 'negative mass' does not exist!) leads to a rapid collapse if thermodynamic equilibrium is to be achieved. This issue will be explored in more detail in chapter 8. However, for the rest of the stellar systems, especially galaxies, we can assume relaxation is never achieved. This allows us to adopt the approximation that as a star moves in phase space, it is affected only by the (long-range) gravitational potential of the system, and not by local interactions. *Such behaviour represents one of the fundamental properties of dynamical systems moving under gravitation.*

3.4 Local and distant encounters

We can gain more insight into the role that local and distant encounters play in relaxation by giving a more realistic description of a two-body interaction. In section 2.2 we derived the trajectories in a Kepler potential, finding them to be conic sections. We will focus here on scattering events, tracing hyperbolic trajectories as illustrated in figure 3.2 for the interaction between two stars with masses m and M. D is the impact parameter – the closest approach had there been no gravitational attraction – and v_∞ is the original velocity of incoming star m (we will use the rest frame of star S_2). As $r \to \infty$, the azimuthal angle $\phi \to \pi/2 + \theta/2$.

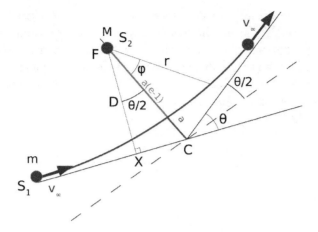

Figure 3.2 Computing the relaxation time: encounter of two stars on an unbound hyperbolic trajectory.

Hence

$$1 + e\cos(\pi/2 + \phi/2) = 0 \qquad \Rightarrow \qquad \sin(\theta/2) = \frac{1}{e}. \qquad (3.15)$$

Using this expression along with the derivation of the angle $\theta/2$ from the triangle $X\hat{F}C$, we get

$$\cot(\theta/2) = \frac{D}{a}. \qquad (3.16)$$

Equations 3.15 and 3.16 can also be used to find the eccentricity:

$$e = \sqrt{1 + \left(\frac{D}{a}\right)^2} = \sqrt{\frac{2}{1 - \cos\theta}}. \qquad (3.17)$$

When $D \to a$, $\theta \to \pi/2$. Another useful equation here is the *vis viva* integral found in the Kepler problem. Along the path of the orbit (which, in general, could correspond to any conic section), the quantity

$$v^2 = G(M + m)\left(\frac{2}{r} \pm \frac{1}{a}\right) \qquad (3.18)$$

is conserved (use the negative sign for an elliptical orbit, and the positive sign for a hyperbola). In the case $r \to \infty$,

$$v_\infty^2 = \frac{G(M + m)}{a} \qquad \Rightarrow \qquad \cot(\theta/2) = \frac{Dv_\infty^2}{G(M + m)}. \qquad (3.19)$$

One interesting result from this expression is the concept of 'gravitational focusing'. Let us compare the cross section (i.e., the area within which a strong gravitational interaction is likely) with the geometric area of the stars (i.e., πr_1^2 and πr_2^2, where r_1 and r_2 are the radii of stars S_1 and S_2, respectively). A physical collision will take place if the true point of closest approach is smaller than the sum of the stellar radii, i.e., $a(e-1) \leq r_1 + r_2$. From (3.18)

$$\left(\frac{v_{\text{MAX}}}{v_\infty}\right)^2 = 1 + \frac{2a}{r_{\min}} = \frac{r_{\min}(2a + r_{\min})}{r_{\min}^2} \tag{3.20}$$

leads to

$$\frac{v_{\text{MAX}}}{v_\infty} = \frac{D}{r_1 + r_2}. \tag{3.21}$$

For typical values in the solar neighbourhood ($v_{\text{MAX}} \sim v_{\text{esc}} \sim 620\,\text{km s}^{-1}$; $v_\infty \sim 30\,\text{km s}^{-1}$) we get a difference between the actual cross section of interaction and the simple geometric cross section of

$$\left(\frac{D}{r_1 + r_2}\right)^2 \sim 400, \tag{3.22}$$

leading to the idea of focusing: two stars will get closer because of their mutual gravitational interaction, increasing their chances of collision by a factor ~ 400 with respect to a simple geometric argument (i.e., no long-range forces). Taking the local stellar mass density in the solar neighbourhood as reference, $\rho_0 \sim 0.1 M_\odot \text{pc}^{-3}$, we obtain a typical separation between stars $\lambda \sim 2\,\text{pc}$. Given that $\lambda \sim D$, we find that $r_{\text{MIN}} \sim 2/20\,\text{pc} \gg r_1 + r_2$. Therefore, even if focusing is taken into account, the probability of physical collisions is negligible. Only in the densest systems, such as the cores of globular clusters, we can expect this type of effect to be relevant.

Let us now revisit the issue of relaxation time (section 3.3). We could adopt the more rigorous treatment of hyperbolic orbits to deal with the two-body interaction, and follow a similar treatment as in our rectilinear case above, integrating with respect to the parameters of the collision to create an averaged version of the process. The method is a bit more involved, but follows the same argument, estimating a parameter that traces the interactions as a stochastic process that introduces a random walk in the evolution of the kinetic energy of a test star. We refer the reader to the standard treatment of Chandrasekhar,[1] and show the final result for the relaxation time:

$$\tau_{\text{REL}} = \frac{1}{16} \sqrt{\frac{3}{\pi}} \frac{\langle v^2 \rangle^{3/2}}{nG^2 m^2 \ln\left(Rv^2/GM\right)}, \tag{3.23}$$

where v and m are the typical velocities and masses of stars in the system, and n gives their number density. R is the upper limit to the impact parameter, similar to the derivation followed for the rectilinear case, for which we assume this term to be roughly comparable to the size of the whole system. Just to give a simple estimate, the solar neighbourhood $(m \sim 1M_\odot, v \sim 30\,\mathrm{km\,s^{-1}}, n \sim 0.1\,\mathrm{pc^{-3}}, R \sim 300\,\mathrm{pc})$ has a relaxation time $\tau_{\mathrm{REL}} \sim 5 \times 10^{13}\,\mathrm{yr}$, therefore much longer than the Hubble time. This treatment allows us to determine the contribution to the potential relaxation process from nearby or distant 'collisions': here we define a 'collision' as a gravitational interaction that will produce some deflection of the trajectory of the test star. From equation 3.19 and noting that v_∞ represents the typical velocity of a test star, we can write the Coulomb logarithm as $\ln(R/a)$, where a is defined by the hyperbolic trajectory of the star. We can now assess the contribution to the relaxation time from collisions with an impact parameter D by calculating the ratio

$$X(D) \equiv \frac{\ln(D/a)}{\ln(R/a)}, \tag{3.24}$$

where D/a is related to the deflection angle θ, via equation 3.16. Let us consider an example at the highest stellar densities, say, a globular cluster core $(a \sim 9 \times 10^{-5}\,\mathrm{pc}; R = r_{\mathrm{core}} \sim 0.3\,\mathrm{pc})$. In this scenario, collisions with $D/a \sim 1.5$ lead to $\theta \sim 67°$ (i.e., very large deflections), and the contribution is just $X \sim 5\%$. At weaker deflection angles, say $D/a \sim 10$ (corresponding to $\theta \sim 10°$), the contribution increases to $X \sim 30\%$, and for very weak scattering events, i.e., those caused by far away stars (say $D/a \sim 500 \longrightarrow \theta \sim 0.2°$), the contribution is $X \sim 75\%$. Therefore, we conclude that in galaxy dynamics, most of the 'collisions' between stars that eventually cause relaxation are distant.

3.5 Collisionless Boltzmann equation

Given that in most circumstances a stellar system can be treated as an ensemble of particles without any collisions, with individual trajectories corresponding to the underlying gravitational potential, we can describe this motion in phase space according to a simple differential equation. Consider the motion of these stars in phase space as some sort of 'fluid', moving through a differential element of phase space, $d\tau \equiv d^3r\,d^3v$. We will use Cartesian coordinates for simplicity, so that the differential element is a six-dimensional hypercube. We want to describe the change in phase space density, $f(\vec{r}, \vec{v}; t)$, following two steps. First we write down the flux of stars through the opposite faces in position of the hypercube. For instance, through the faces normal to the $\hat{e}_x$ direction, we have a flux difference of

$$\left\{fu - \left[fu + \frac{\partial(fu)}{\partial x}du\right]\right\} dy\,dz\,du\,dv\,dw = -\frac{\partial(fu)}{\partial x}d\tau, \qquad (3.25)$$

which, combining all three spatial directions, leads to the first part of the variation of the distribution function:

$$-\vec{\nabla}_{\vec{r}}(f\vec{v})d\tau = -\vec{v}\cdot\vec{\nabla}_{\vec{r}}f\,d\tau. \qquad (3.26)$$

The second part of the variation of the distribution function arises from the flux through the 'velocity' faces of the hypercube. For instance, the contribution along the $\hat{e}_u$ direction (i.e., the x-component of the velocity) is

$$\left\{f\frac{du}{dt} - \left[f\frac{du}{dt} + \frac{\partial(f\frac{du}{dt})}{\partial u}du\right]\right\} dy\,dz\,du\,dv\,dw = -\frac{\partial(f\frac{du}{dt})}{\partial u}d\tau =$$
$$= \frac{\partial f}{\partial u}\frac{\partial \Phi}{\partial x}d\tau. \qquad (3.27)$$

The final expression uses the fact that the x-component of the acceleration can be written with respect to the gradient of the gravitational potential (from chapter 2). Note that this step also needs to assume the potential does not depend on the velocity (always true in Newtonian gravity). The variation of the number of stars within this volume of phase space should equal the implicit dependence of the distribution function with time, $\partial_t f$. The combination of all these terms results in a continuity equation:

$$\frac{\partial f}{\partial t} + \vec{v}\cdot\vec{\nabla}_{\vec{r}}f - \vec{\nabla}\Phi(\vec{r})\cdot\vec{\nabla}_{\vec{v}}f = 0. \qquad (3.28)$$

The above equation is the collisionless Boltzmann equation (hereafter CBE). We can understand this equation if we consider that in fluid dynamics, the Lagrangian derivative is defined as the rate of change of a physical quantity in the fluid as one follows its trajectory. In an ordinary fluid, this derivative, applied to the fluid density (ρ) can be written

$$D\rho = \partial_t\rho + \vec{v}\cdot\vec{\nabla}\rho = 0. \qquad (3.29)$$

We can extend this concept to phase space, so that the CBE can be written with respect to a Lagrangian derivative as

$$Df = 0. \qquad (3.30)$$

Therefore, the motion of stars in phase space is such that as we follow the trajectory of one of these stars, we do not find any local variation of the stellar number density (Liouville's theorem). An alternative way to arrive at this interpretation is by looking at the CBE as an equation in partial derivatives in seven coordinates (phase space plus time), leading to six subsidiary, first order, ordinary differential equations:

$$\frac{dt}{1} = \frac{dx}{u} = \frac{dy}{v} = \frac{dz}{w} = \frac{du}{\left(-\frac{\partial \Phi}{\partial x}\right)} = \frac{dv}{\left(-\frac{\partial \Phi}{\partial y}\right)} = \frac{dw}{\left(-\frac{\partial \Phi}{\partial z}\right)}. \qquad (3.31)$$

These correspond to the standard equations of motion of a test particle under the gravitational potential of the system.

Exercise 3.2

Give the *two* main reasons why the distribution of stars in the Milky Way galaxy cannot be described by analogy with respect to the distribution of air molecules in a room. How about the description of stars in a dense globular cluster?

3.6 Isolating integrals: Jeans theorem

The CBE governs the behaviour of the distribution function in phase space. It encodes all information about the dynamical system, although normally we use the velocity moments to make comparisons between models and observations. It is not trivial to determine the distribution function that corresponds to a stellar system, and the use of conserved quantities is fundamental to be able to operate with this formalism. In this context, we can separate conserved quantities in two categories: isolating and nonisolating integrals of motion. The best way to visualize the difference between them is to take pairs of dynamical variables (i.e., position and its associated momentum) as action-angle pairs. It is easiest to picture the evolution of such a pair: a trajectory will cover the surface of a torus (see figure 3.3). The trajectory of a star with a given constant of motion may be as in panel (a) of the figure, where the orbit is closed and the whole trajectory does not cover the whole surface of the torus. Another option is shown in panel (b), where the trajectory does

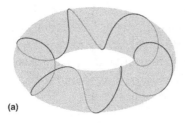 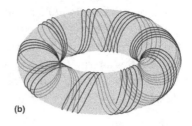

(a)   (b)

Figure 3.3 Illustration of the difference between nonisolating (a) and isolating (b) integrals.

not close and ergodically covers the whole surface. Keeping in mind that no two orbits should cross (as they correspond to different integrals of motion), one could say that by choosing a conserved quantity that behaves as in panel (b), we can separate this part of phase space volume from the rest, by choosing that specific value of the integral. Different values of the integral will define different tori. A choice of the related constant of motion can thus be used to 'compartmentalize' phase space. These constants of motion are called 'isolating integrals'. In the more general case, a given isolating integral will divide phase space into disjoint hypersurfaces, a second isolating integral will define a different set of hypersurfaces, and so on, so that one could fully describe the distribution function with respect to these integrals, rather than the standard way using position and velocity (Jeans theorem). Moreover, these isolating integrals emanate from a symmetry of the system, allowing us to produce educated guesses for the distribution function of a stellar system, as we will see in section 3.7.

The isolating integrals most often found in stellar systems are energy, expected in stationary systems, i.e., those without an explicit time dependence in the distribution function; total angular momentum (and its components) if the system is spherically symmetric; or a single component of angular momentum if only rotational symmetry is present about one of the axes. For instance, in the Milky Way galaxy, to lowest approximation, we may consider the system to be in steady state and with cylindrical symmetry, so that E (energy) and J_z (the projection of angular momentum along the axis of rotation) are isolating integrals.[2] In chapter 4 we will see that the motion of stars in the Galaxy suggests an additional third integral. Although finding isolating integrals is non trivial, one can make use of the properties of the system to produce educated guesses.

3.7 Examples of distribution functions

We will show here the properties of some distribution functions (DFs) frequently used to describe stellar systems. All systems are assumed to be in a stationary state. Hence, energy (E) is always an isolating integral. Note that if E is the only isolating integral, the velocity distribution must be isotropic. This stems from the fact that $E = v^2/2 + \Phi(\vec{r})$, so that the second order moment of the j-th component of velocity is

$$\langle v_j^2(\vec{r}) \rangle = \frac{1}{\nu(\vec{r})} \iiint v_j^2 F(E) d^3v, \tag{3.32}$$

leading to $\langle v_x^2(\vec{r}) \rangle = \langle v_y^2(\vec{r}) \rangle = \langle v_z^2(\vec{r}) \rangle$. Therefore, if the observed velocity distribution is not isotropic, there must be an additional isolating integral that breaks this symmetry. Such is the case with motion in an axially symmetric system (e.g., a rotating disc), where the projection of angular momentum along the symmetry axis (say $J_z = rv_\perp$) breaks the symmetry with respect to v^2.

When defining distribution functions, one commonly encounters an energy floor, corresponding to the minimum of the effective potential (Φ_0), from which a relative potential can be defined as

$$\Psi(\vec{r}) \equiv -\Phi(\vec{r}) + \Phi_0, \tag{3.33}$$

as well as a relative energy:

$$\mathcal{E}(\vec{r}) \equiv -\frac{v^2}{2} + \Psi(\vec{r}). \tag{3.34}$$

Polytrope

The polytrope model defines the distribution function as some power law of the relative energy:

$$F(\mathcal{E}) = \begin{cases} A\mathcal{E}^{n-\frac{3}{2}}, & \mathcal{E} > 0 \\ 0, & \mathcal{E} \leq 0, \end{cases} \tag{3.35}$$

where n is the polytropic index. The density can be written with respect to the relative potential by integrating in velocity space:

$$\rho = m \iiint F(\mathcal{E}) d^3v = 4\pi m \int_0^{\sqrt{2\Psi}} F(\Psi - \frac{v^2}{2}) v^2 dv = \cdots = c_n \Psi^n, \tag{3.36}$$

where the coefficient c_n is defined as

$$c_n = mA(2\pi)^{3/2} \frac{(n - \frac{3}{2})!}{n!}.$$ (3.37)

Note that for this coefficient to converge, we need $n > 1/2$. Therefore the case $n = 0$ (i.e., a homogeneous polytrope) is not physically possible. We can write the equation that leads to the radial dependence of the density by using Poisson's equations in spherical polar coordinates:

$$\nabla^2 \Psi = -4\pi G\rho \Longrightarrow \frac{1}{r^2} \frac{d}{dr} \left(r^2 \frac{d\Psi}{dr} \right) + 4\pi G c_n \Psi^n = 0,$$ (3.38)

which is the Lane-Emden equation, typically found in the analysis of stellar structure, when describing a system in hydrostatic equilibrium, as illustrated in exercise 3.3.

Exercise 3.3

By assuming hydrostatic equilibrium in a spherical distribution of gas with a polytropic equation of state $p \propto \rho^\gamma$, show that we recover the equivalent case to a collisionless polytrope (i.e., equation 3.38), with $\gamma = 1 + 1/n$.

Two interesting cases arise for the polytrope model: $n = 5$ produces the Plummer sphere presented in chapter 2, and $n \to \infty$ leads to the isothermal model, presented below.

Isothermal sphere

In stellar dynamics, the isothermal sphere is the thermodynamic equivalent to a Maxwell-Boltzmann distribution. Replacing the kinetic energy by the total energy, and the temperature (which behaves as an energy scale, kT) by a velocity dispersion parameter (σ), the isothermal sphere is

$$F(\mathcal{E}) = \begin{cases} Ae^{\frac{\mathcal{E}}{\sigma^2}}, & \mathcal{E} > 0 \\ 0, & \mathcal{E} \leq 0. \end{cases}$$ (3.39)

> **Exercise 3.4**
>
> Show that for the isothermal sphere distribution function defined in equation 3.39, the density profile is $\rho(r) \propto 1/r^2$.

The isothermal sphere is one of the most useful distributions in gravitational dynamics. Note that the velocity of a circular orbit is constant (v_0):

$$v^2(r) = \frac{GM(<r)}{r} = \frac{G}{r} \int_0^r 4\pi s^2 \frac{\mathcal{K}}{s^2} ds = 4\pi G\mathcal{K} \equiv v_0^2 \Longrightarrow \rho(r) = \frac{v_0^2}{4\pi Gr^2}. \tag{3.40}$$

Also note that the cumulative mass profile increases linearly with radius. Therefore, in order to avoid a divergent mass, the density is truncated at some radius r_0. This defines the so-called Truncated Singular Isothermal Sphere. In spiral galaxies, the velocity profile outside of the central regions is roughly flat, suggesting a quasi-isothermal distribution.

Mestel disc

Mestel disc[3] is a distribution related to the dynamics in disc galaxies such as our Milky Way. This distribution features a net bulk rotation, and the stars move within a very thin disc. Therefore, the three-dimensional mass density, $\rho(r)$, is replaced by a two-dimensional density, $\Sigma(R)$. Note that, hereafter, we distinguish between the 3D radial distance (r) and the 2D projection (R). The equivalent of an isothermal sphere in a 2D system is given by $\Sigma(R) \propto 1/R$. Denoting v_0 as the radius-independent velocity of any circular orbit in this system, we can write the surface mass density as

$$\Sigma(R) = \frac{v_0^2}{2\pi GR}, \tag{3.41}$$

and the associated relative potential as

$$\Psi(R) = -v_0^2 \ln\left(\frac{R}{R_0}\right), \tag{3.42}$$

where we assume the gravitational potential, Φ, vanishes at $R = R_0$. The cumulative mass profile also increases linearly with radius and calls for a truncated model, where the mass density vanishes at $R > R_0$.

Following Jeans theorem, we assume that energy (E) and the vertical component of angular momentum (J_z) are isolating integrals (we assume the disc is located on the xy horizontal plane). We follow the previous ansatz of an isothermal distribution for the energy and add a power law term with respect to angular momentum:

$$F(\mathcal{E}) = \begin{cases} A \left(\dfrac{J_z}{R_0 v_0} \right)^q e^{\frac{\mathcal{E}}{\sigma^2}}, & J_z > 0 \wedge \mathcal{E} > 0 \\ 0, & J_z \leq 0 \vee \mathcal{E} \leq 0. \end{cases} \tag{3.43}$$

The condition on the sign of J_z eliminates counterrotating stars. From the DF, we can derive the surface mass density:

$$\Sigma(R) = \iint F(\mathcal{E}, J_z) d^2 v =$$

$$= A \int_0^\infty dv_\perp \int_{-\infty}^{+\infty} dv_\parallel \left(\frac{R v_\perp}{R_0 v_0} \right)^q \exp \left[-\frac{v_\parallel^2 + v_\perp^2}{2\sigma^2} - \frac{v_0^2}{\sigma^2} \ln \left(\frac{R}{R_0} \right) \right],$$

$$\tag{3.44}$$

leading to

$$\Sigma(R) = A\sigma^2 \pi^{\frac{1}{2}} 2^{\frac{q}{2}} \Gamma(\frac{q}{2} + \frac{1}{2}) \left(\frac{R\sigma}{R_0 v_0} \right)^q \left(\frac{R}{R_0} \right)^{-v_0^2/\sigma^2}. \tag{3.45}$$

This result is compatible with our original estimate of the surface mass density (equation 3.41) if

$$q = \frac{v_0^2}{\sigma^2} - 1. \tag{3.46}$$

This expression allows us to consider two different regimes: if q is a large number, the system is a cold dynamical disc, where the kinetic energy is mostly in bulk rotational motion ($v_0 \gg \sigma$). If $q \to -1$, then we have a hot dynamical disc, and the kinetic energy budget is dominated by random motion (loosely called 'pressure').

Osipkov-Merritt models

An extension that includes angular momentum in the DF to a spherical system is given by the Osipkov-Merritt models.[4] We consider a 3D system where the isolating integrals are energy ($\mathcal{E}$) and angular momentum (J), but the dependence of the DF is given by a variable Q:

$$f(\vec{r}, \vec{v}) = F(Q), \qquad \text{where } Q \equiv \mathcal{E} + \frac{J^2}{2r_a^2}. \tag{3.47}$$

Note that this expression is motivated by the contribution of angular momentum to the energy. For instance, in the Kepler problem (see chapter 2), the total energy can be written

$$E = \frac{\dot{R}^2}{2} - \frac{GM}{R} + \frac{J^2}{2R^2}.$$ (3.48)

In this case, r_a is a free parameter that relates to the anisotropy of the orbits, as we will see below. We separate the velocity vector into a radial and a tangential component: $\vec{v} = v_r \vec{e}_r + \vec{v}_t$. The mass density is

$$\rho(r) = \iiint F(Q) d^3 v = 2\pi \int dv_r \int dv_t v_t F(Q) = \cdots$$

$$\cdots = \frac{2\pi}{1 + \frac{r^2}{r_a^2}} \int_\Phi^0 dQ \sqrt{2(Q - \Phi)} F(Q),$$ (3.49)

where $\Phi(r)$ is the gravitational potential. We are interested in the anisotropy of the velocity distribution, so let us compute the radial and tangential components of the velocity dispersion tensor:

$$\rho \langle v_r^2 \rangle = \cdots = \frac{1}{3} \frac{2\pi}{1 + \frac{r^2}{r_a^2}} \int_\Phi^0 dQ \, [2(Q - \Phi)]^{3/2} F(Q),$$

$$\rho \langle v_t^2 \rangle = \cdots = \frac{2}{3} \frac{2\pi}{1 + \frac{r^2}{r_a^2}} \int_\Phi^0 dQ \, [2(Q - \Phi)]^{3/2} F(Q).$$ (3.50)

Therefore, the level of anisotropy is given by the 'beta' parameter:

$$\beta \equiv 1 - \frac{\sigma_t^2}{\sigma_r^2} = \frac{r^2}{r^2 + r_a^2}.$$ (3.51)

In this model, the central regions ($r \ll r_a$) have an isotropic velocity distribution ($\beta \sim 0$), whereas the outskirts ($r \gg r_a$) feature radial anisotropy ($\beta \sim 1$). This behaviour is typical of galaxies, where the outer regions have a more significant presence of stars on orbits which are very close to radial trajectories, reflecting events in the mass assembly history. Remember that in galaxy dynamics, the relaxation times are very long; hence stellar orbits retain their past dynamical history for long periods of time.

3.8 Jeans equations

The CBE is defined in the six-dimensional phase space, making it hard to exploit with observational data, which offer instead an 'integrated' mapping of this equation. The Jeans equations provide a more tractable set of equations for comparisons with data. They consist of moments of velocity:

$$\text{Jeans equations} \Longrightarrow \int d^3 v v_i^n \, (\text{CBE}) , \tag{3.52}$$

where n is the moment order. To avoid overloading the expressions in the rest of this chapter, we will use a single integral sign to denote 'velocity-volume' integrals. Let us begin with the zeroth order moment:

$$\int \partial_t f d^3 v + \sum_i \int v_i \partial_{x_i} f d^3 v - \sum_i \partial_{x_i} \Phi \int \partial_{v_i} f d^3 v = 0. \tag{3.53}$$

The last term vanishes: apply the divergence theorem, which reduces the volume integral to a surface integral evaluated at $v \to \pm \infty$, where the distribution function should vanish. Note we have also pulled the spatial and time dependence out of the velocity integrals as these are independent variables. The remaining integrals refer to the number density $v(\vec{r}; t)$ and average velocity $\langle v_i(\vec{r}; t) \rangle$, leading to a continuity equation, this time on averaged variables:

$$\partial_t v + \sum_i \partial_{x_i} (v \langle v_i \rangle) = 0. \tag{3.54}$$

We can relate the stellar number density, v, to the mass density in stars, $\rho = mv$, where m is the mass of a single star.

The first order moment of the j-th component of velocity is obtained in an analogous way:

$$\int \partial_t f v_j d^3 v + \sum_i \int v_i v_j \partial_{x_i} f d^3 v - \sum_i \partial_{x_i} \Phi \int v_j \partial_{v_i} f d^3 v = 0. \tag{3.55}$$

Applying again the divergence theorem to the last term and assuming a vanishing surface integral (in velocity space) leads to the following:

$$\partial_t (v \langle v_j \rangle) + \sum_i \partial_{x_i} (v \langle v_i v_j \rangle) + v \partial_{x_j} \Phi = 0. \tag{3.56}$$

The second term is usually written with respect to the covariance of the velocity distribution (the velocity dispersion tensor):

$$\sigma_{ij}^2 \equiv \langle (v_i - \langle v_i \rangle)(v_j - \langle v_j \rangle) \rangle = \langle v_i v_j \rangle - \langle v_i \rangle \langle v_j \rangle, \tag{3.57}$$

leading to

$$v \partial_t \langle v_j \rangle + v \langle \vec{v} \rangle \cdot \vec{\nabla}_{x_i} \langle v_j \rangle + v \partial_{x_j} \Phi + \sum_i \partial_{x_i} (v \sigma_{ij}^2) = 0. \tag{3.58}$$

Jeans equations in spherical coordinates

We will show below the derivation of a very useful representation of Jeans equations. We assume here spherical symmetry, and consider only the equation created by multiplying the CBE by the radial component of the velocity. The three components of velocity and acceleration in spherical coordinates are shown in the equation in exercise 3.5.

Exercise 3.5

Show that in spherical polar coordinates (r,θ,ϕ), where θ and ϕ are the polar and azimuthal angles, respectively, the three components of velocity are $v_r = \dot{r}$; $v_\theta = r\dot{\theta}$ and $v_\phi = r\dot{\phi}\sin\theta$, and the three components of the acceleration are

$$\left. \begin{aligned} a_r &= \ddot{r} - r\dot{\theta}^2 - r\dot{\phi}^2 \sin^2\theta \\ a_\theta &= r\ddot{\theta} + 2\dot{r}\dot{\theta} - r\dot{\phi}^2 \sin\theta\cos\theta \\ a_\phi &= r\ddot{\phi}\sin\theta + 2\dot{r}\dot{\phi}\sin\theta + 2r\dot{\phi}\cos\theta \end{aligned} \right\}.$$

(Hint: find first the time derivatives of the unit vectors: $\dot{\hat{e}}_r$, $\dot{\hat{e}}_\theta$, $\dot{\hat{e}}_\phi$.)

We begin with the most general expression for the CBE:

$$\partial_t f + \sum_i \dot{r}_i \partial_{r_i} f + \sum_i \dot{v}_i \partial_{v_i} f = 0, \tag{3.59}$$

and use the vector equation $\dot{\vec{v}} = -\vec{\nabla}\Phi$ in spherical polar coordinates, noting that, by the assumed symmetry, the potential depends only on the radial component, leading to

$$
\left.
\begin{aligned}
\dot{v}_r &= -\frac{d\Phi}{dr} + \frac{v_\theta^2 + v_\phi^2}{r} \\[2mm]
\dot{v}_\theta &= \frac{v_\phi^2 \cot\theta - v_r v_\theta}{r} \\[2mm]
\dot{v}_\phi &= -\frac{v_\phi v_r + v_\phi v_\theta \cot\theta}{r}
\end{aligned}
\right\}.
\tag{3.60}
$$

We replace these expressions into the CBE, and follow the same procedure as above, multiplying by the radial component of the velocity and integrating in all of velocity space. Let us write the expression in full:

$$
\partial_t \int d^3 v\, v_r f + \partial_r \int d^3 v\, v_r^2 f + \partial_\theta \int d^3 v\, v_r v_\theta f + \partial_\phi \int d^3 v\, v_r v_\phi f
$$

$$
+ \int d^3 v\, v_r \left(\frac{v_\theta^2 + v_\phi^2}{r} - \frac{d\Phi}{dr} \right) \partial_{v_r} f + \int d^3 v\, v_r \frac{v_\phi^2 \cot\theta - v_r v_\theta}{r} \partial_{v_\theta} f
$$

$$
- \int d^3 v\, v_r \frac{v_\phi v_r + v_\phi v_\theta \cot\theta}{r} \partial_{v_\phi} f = 0.
\tag{3.61}
$$

Some of the integrals can be readily evaluated as averages of velocities (using equations 3.2, 3.3, 3.4). The last three terms require integration by parts, where we are left with integrals such as

$$
\int d^3 v\, g(v_i, \cdots) \partial_{v_i} f = [f g(v_i, \cdots)]_{-\infty}^{+\infty} - \int d^3 v\, f \partial_{v_i} g(v_i, \cdots),
\tag{3.62}
$$

and the first term vanishes, as it is evaluated in the limit towards infinite velocities, where we expect the distribution function to decrease faster than the increase in any other velocity-related expression – to avoid, for instance, an infinite velocity average or velocity dispersion. The expression simplifies to

$$
\partial_t(v \langle v_r \rangle) + \partial_r(v \langle v_r^2 \rangle) + \partial_\theta(v \langle v_r v_\theta \rangle) + \partial_\phi(v \langle v_r v_\phi \rangle)
$$

$$
- v \left(\frac{\langle v_\theta^2 \rangle + \langle v_\phi^2 \rangle}{r} - \frac{d\Phi}{dr} \right) + 2v \frac{\langle v_r^2 \rangle}{r} = 0
\tag{3.63}
$$

Since the system is spherically symmetric, we expect the average velocities to vanish ($\langle v_i \rangle = 0$), allowing us to write $\sigma_i^2 = \langle v_i^2 \rangle$. The velocity dispersion tensor can have only two non-zero components: $\sigma_{rr} \equiv \sigma_r$ and $\sigma_{\theta\theta} = \sigma_{\phi\phi} \equiv \sigma_\perp$. The off-diagonal terms vanish. Therefore, the final equation is

$$\frac{1}{\nu}\frac{\partial(\nu\sigma_r^2)}{\partial r} + \frac{2\sigma_r^2}{r}\beta + \frac{d\Phi}{dr} = 0, \qquad (3.64)$$

where we have defined the anisotropy parameter $\beta \equiv 1 - (\sigma_\perp^2/\sigma_r^2)$. We will also see in chapter 4 the equivalent version of Jeans equations in cylindrical symmetry.

Exercise 3.6

Let us describe a dark matter halo as a spherically symmetric distribution with a constant and isotropic velocity dispersion σ. The halo includes a baryon, diffuse, gaseous component without a temperature gradient. The dark matter density dominates the mass budget ($\rho_{DM} \gg \rho_g$). With the use of Jeans equation, assuming a stationary state and no bulk motion, show that the gas density profile and the dark matter density profile are related via:

$$\rho_g(r) \propto [\rho_{DM}(r)]^\eta, \qquad \text{where } \eta = \frac{\mu m_p \sigma^2}{kT}.$$

If the anisotropy parameter, β, is a function of the radial coordinate, the solution to the homogeneous version of equation 3.64 is

$$\left(\nu\sigma_r^2\right) = \mathcal{K}\exp\left(-2\int_0^r \frac{\beta(p)}{p}dp\right), \qquad (3.65)$$

and $\mathcal{K}$ is a constant that trivially corresponds to the value of $(\nu\sigma_r^2)$ at the origin of coordinates. The general solution can be obtained by promoting the constant to a radial function $\mathcal{K}(r)$, leading to

$$\left(\nu\sigma_r^2\right) = \int_r^\infty \left[\frac{GM(<s)\nu(s)}{s^2}\exp\left(2\int_r^s \frac{\beta(p)}{p}dp\right)\right]ds, \qquad (3.66)$$

where we have applied the boundary condition $\mathcal{K}(r) \to 0$ as $r \to \infty$.

3.9 The virial theorem

The Jeans equations were obtained by removing information about individual velocities and replacing them by averaged quantities. Furthermore, one can integrate the Jeans equations over the spatial coordinates, effectively removing all space-related information and leading to an equation that deals with the energy balance in the dynamical system. This is the virial theorem:

$$\text{virial theorem} \Longrightarrow \int d^3x x_k (\text{Jeans equations}). \qquad (3.67)$$

Let us start from the equation corresponding to the j-th component of velocity (equation 3.56):

$$\int x_k \partial_t (\rho \langle v_j \rangle) d^3x = -\sum_i \int x_k \partial_{x_i} \left(\rho \langle v_i v_j \rangle \right) d^3x - \int \rho x_k \partial_{x_j} \Phi d^3x =$$

$$= \sum_i \int \rho \langle v_i v_j \rangle \delta_{ik} d^3x + \mathcal{W}_{kj} = 2\mathcal{K}_{kj} + \mathcal{W}_{kj}, \qquad (3.68)$$

where we have integrated by parts the first term on the RHS and applied the divergence theorem, so that the surface integral vanishes as $x \to \infty$. The two integrals on the RHS are symmetric with respect to the indices k, j, and we define them as $\mathcal{K}$ and $\mathcal{W}$, representing the kinetic and potential energy tensors, respectively. The kinetic energy tensor can be split into two parts related to bulk motion ($\mathcal{T}$) and bulk motion (Π):

$$\mathcal{K}_{kj} = \frac{1}{2} \int \rho \langle v_k v_j \rangle d^3x = \frac{1}{2} \int \rho \left[\langle v_k \rangle \langle v_j \rangle + \sigma_{kj}^2 \right] d^3x \equiv \mathcal{T}_{kj} + \Pi_{kj}. \quad (3.69)$$

The gravitational potential energy tensor can be duly identified if we write it as

$$\mathcal{W}_{kj} = G \int \rho(\vec{x}) x_k \frac{\partial}{\partial x_j} \left[\int \frac{\rho(\vec{x}')}{|\vec{x} - \vec{x}'|} d^3x' \right] d^3x =$$

$$= G \iint \rho(\vec{x}) \rho(\vec{x}') \frac{x_k (x_j' - x_j)}{|\vec{x} - \vec{x}'|^3} d^3x d^3x'. \qquad (3.70)$$

The expression is symmetric with respect to the integrating variables $\vec{x}$ and $\vec{x}'$. We can symmetrize it, i.e., add a new term switching these variables, and multiply all by a factor of $\frac{1}{2}$, to get

$$W_{kj} = -\frac{G}{2} \iint \rho(\vec{x})\rho(\vec{x}')\frac{(x'_k - x_k)(x'_j - x_j)}{|\vec{x} - \vec{x}'|^3} d^3x d^3x'. \tag{3.71}$$

The trace of this expression is

$$\mathrm{tr}\, W = \sum_{j=1}^{3} W_{jj} = \frac{1}{2} \int \rho(\vec{x})\Phi(\vec{x})d^3x, \tag{3.72}$$

which is the standard gravitational potential energy (see equation 2.23).

Exercise 3.7

Elliptical galaxies are believed to form from the mergers of progenitor systems that can be either discs or other ellipticals. Let us consider two elliptical galaxies with the same mass (M), size (r_0) and velocity dispersion (σ_0), slowly approaching from infinity and eventually merging to form a more massive galaxy that settles into virial equilibrium. Show that the total energy of the system is $E_{\mathrm{TOT}} = -3M\sigma_0^2$. Find the velocity dispersion, size and density of the new galaxy, assuming that its mass distribution is identical to that of the original galaxies.

3.10 Beyond the collisionless Boltzmann equation: The Fokker-Planck equation

When collisions need to be taken into account (e.g., when describing the dense, central region of a globular cluster), the CBE is not an adequate description of the phase space density, and an additional term needs to be considered:

$$\frac{Df}{dt} = \frac{\partial f}{\partial t} + \vec{v}\cdot\vec{\nabla}_r f - \vec{\nabla}_r\Phi\cdot\vec{\nabla}_v f = \left(\frac{\partial f}{\partial t}\right)_{\mathrm{coll}}. \tag{3.73}$$

The term on the RHS takes into account the change in phase space density from collisions. We can describe this term by a probability distribution function, such that $p(\vec{v}, \Delta\vec{v})d\Delta\vec{v}$ gives the probability that a particle with velocity $\vec{v}$ experiences a shift in velocity $\Delta\vec{v}$ within a differential element $d\Delta\vec{v}$ over a time interval Δt. This probability gives information about individual encounters. According to this definition, we can write

$$f(\vec{v}, t + \Delta t) = \int_{-\infty}^{+\infty} f(\vec{v} - \Delta\vec{v}, t) p(\vec{v} - \Delta\vec{v}, \Delta\vec{v}) d^3\Delta v, \qquad (3.74)$$

thereby transforming the Boltzmann equation into a rather complicated integro-differential equation (called the 'master equation').

Even in dense stellar systems, we are still in a regime where the relaxation time is long enough. Therefore $\Delta v \ll v$ during a crossing time, and we can perform a Taylor expansion (LHS in Δt and RHS in $\Delta\vec{v}$) and truncate at the second order; this is the Fokker-Planck approximation:

$$\left(\frac{\partial f}{\partial t}\right)_{coll} = -\sum_{i=1}^{3} \frac{\partial (f \langle \Delta v_i \rangle)}{\partial v_i} + \frac{1}{2}\sum_{i=1}^{3}\sum_{j=1}^{3} \frac{\partial^2 (f \langle \Delta v_i \Delta v_j \rangle)}{\partial v_i \partial v_j}, \qquad (3.75)$$

with the first and second order moments of the change in velocity given by

$$\left. \begin{aligned} \langle \Delta v_i \rangle &\equiv \int p(\vec{v}, \Delta\vec{v}) \Delta v_i d^3\Delta v, \\[2mm] \langle \Delta v_i \Delta v_j \rangle &\equiv \int p(\vec{v}, \Delta\vec{v}) \Delta v_i \Delta v_j d^3\Delta v, \end{aligned} \right\} , \qquad (3.76)$$

commonly known as encounter integrals, or diffusion coefficients. The first term in the RHS of equation 3.75 describes a drag process (dynamical friction). The second term describes a diffusion process in velocity space (compare it with the diffusion equation $\partial_t f = -k\nabla^2 f$). Note that the more general treatment of the Fokker-Planck equation (3.73 and 3.75) would involve the diffusion coefficients to be integrated over velocity *and* position. However, we apply the local approximation, where the individual interactions take place over regions much smaller than the size of the system. Hence, we can simplify the treatment of interactions as a superposition of Keplerian hyperbolæ unaffected by the global potential of the system.

Dynamical friction

In equation 3.75 we came across the dynamical friction term. Physically, one can think of it as a drag force (i.e., velocity-dependent) where the passage of a mass, say an object like a black hole, a globular cluster or an incoming galaxy, affects the surrounding region, altering the potential that, in turn, produces a back reaction. It can be represented pictorially as a wake behind the moving mass that changes the local density of stars, introducing an additional force. There is a useful expression to describe dynamical friction, due to Chandrasekhar.[5] We assume that a mass M moves through a homogeneous sea of stars, each of mass $m < M$. The motion can be simplified as a combination of hyperbolic trajectories in a Kepler potential (see section 2.2). Section 3.4 describes the details of a hyperbolic trajectory, but we neglected there the effect of the scattering event on mass M. By considering conservation of momentum, we can relate the reaction on M from the scattering of m as imparting a velocity change:

$$\Delta v_M = - \left(\frac{m}{M+m} \right) \Delta v_m, \qquad (3.77)$$

and we know from section 3.4 that the effect on m is a rotation of the incoming velocity vector $\vec{v}_\infty$ by an angle θ. Now we use the frame of reference of star M, and add many such scattering events as it traverses the field of stars with mass m. For a single scattering event, the net change in velocity is

$$\left. \begin{aligned} \Delta v_\perp &= -\frac{m v_\infty}{M+m} \sin \theta \\[2mm] \Delta v_\parallel &= -\frac{m v_\infty}{M+m} (1 - \cos \theta) \end{aligned} \right\}, \qquad (3.78)$$

where we split the contribution into a parallel and a perpendicular component, with respect to the motion of M. The symmetry of the problem will give, on average, zero net change in $\Delta v_\perp$, and so we need only to consider the parallel contribution. The deflection angle can be written

$$\cos \theta = 1 - \frac{2}{e^2}, \qquad (3.79)$$

where e is the eccentricity of the hyperbolic trajectory. We also introduce here the impact parameter, i.e., the distance of closest approach during the scattering event $- b = a(e - 1)$ – and use the expression of eccentricity from equation 3.17, leading to

$$\Delta v_\parallel = \frac{2m v_\infty}{M+m} \left[1 + \left(\frac{D}{a} \right)^2 \right]^{-1} = \frac{2m v_\infty}{M+m} \left[1 + \frac{b^2 v_\infty^4}{G^2 (M+m)^2} \right]^{-1}. \qquad (3.80)$$

Now we can add up all the contributions from each star as an integral with respect to the impact parameter b, including the number density of stars in this homogeneous sea, by use of the distribution function, that, by definition, can depend only on velocity, $f(\vec{v}_m)$:

$$\frac{d\vec{v}_M}{dt} = (\vec{v}_m - \vec{v}_M) \int_0^{b_{MAX}} 2\pi b \Delta v_\parallel(b) db f(\vec{v}_m) d^3 v_m. \tag{3.81}$$

In this expression, we have replaced $\vec{v}_\infty$ by the *relative* velocity between mass M and one of the stars, i.e., $\vec{v}_m - \vec{v}_M$, so that we can add up the individual contributions. The range of integration in b extends to some maximum value b_{MAX}, normally represented by the size of the stellar distribution. Since the integral produces a logarithmic term (the Coulomb logarithm), our choice is not critical in the derivation:

$$\frac{d\vec{v}_M}{dt} = 4\pi G^2(M+m)m \frac{\vec{v}_m - \vec{v}_M}{|\vec{v}_m - \vec{v}_M|^3} \ln \Lambda f(\vec{v}_m) d^3 v_m, \tag{3.82}$$

where $\Lambda \equiv b_{MAX} v_\infty / G(M+m)$, and we have simplified the Coulomb logarithm by assuming $\ln(1 + \Lambda^2) \approx 2 \ln \Lambda$. This expression corresponds to a specific choice of velocity of the sea of stars. Therefore, we need to integrate it in $\vec{v}_m$. By use of the vector expression encountered in equation 2.11, we can write

$$\int d^3 v_m f(\vec{v}_m) \frac{\vec{v}_m - \vec{v}_M}{|\vec{v}_m - \vec{v}_M|^3} = \vec{\nabla}_{v_M} H(\vec{v}_M), \tag{3.83}$$

and H is called the first Rosenbluth potential. If we assume this potential to be isotropic, the gradient can be written:

$$\vec{\nabla} H = (\partial_{v_M} H) \vec{v}_M / v_M, \tag{3.84}$$

arriving at the final expression for the dynamical friction effect, namely:

$$\frac{d\vec{v}_M}{dt} = -16\pi^2 G^2(M+m)m \ln \Lambda \frac{\vec{v}_M}{v_M^3} \int_0^{v_M} f(v_m) v_m^2 dv_m. \tag{3.85}$$

The integral represents the number density of stars with speeds below v_M. The force is directed against the motion of mass M and depends on its velocity. If we assume $M \gg m$, and a Maxwellian distribution of velocities for the sea of stars, with dispersion σ, we find

$$\frac{d\vec{v}_M}{dt} = -4\pi G^2 M \rho \ln \Lambda \frac{\vec{v}_M}{v_M^3} g(s), \tag{3.86}$$

where ρ is the mass density of stars, $s \equiv v_M/\sigma\sqrt{2}$, and the function $g(s)$ can be written with respect to the error function:

$$g(s) = \mathrm{erf}(s) - \frac{2s}{\sqrt{\pi}} e^{-s^2}. \tag{3.87}$$

Dynamical friction is responsible, for instance, for the orbital decay of globular clusters as they move around the galaxy. Note that this expression assumes a point mass M moving through a homogeneous sea of stars with mass m, but a similar result can be derived for more realistic scenarios, such as the merging of two galaxies. From equation 3.86 we expect a drag force proportional to M^2 and inversely proportional to v_M^2. Therefore, dynamical friction is more efficient when the incoming galaxy in a merger is more massive: major mergers – those where the masses of both galaxies are comparable – proceed faster than minor mergers, but fast encounters – those in high-density environments – will imply a weaker contribution from dynamical friction.

Notes

1 Chandrasekhar, 1960, *Principles of stellar dynamics*, Dover, p. 48.
2 Throughout this book, unless explicitely noted, these isolating integrals are defined per unit mass.
3 This is a simplified version of the model presented in Mestel, 1963, MNRAS, 126, 553.
4 Merritt, 1985, AJ, 90, 1027.
5 Chandrasekhar, 1943, ApJ, 97, 255.

4

Understanding our Galaxy

The Milky Way Galaxy can be described, to the lowest order, as a system of stars in a thin, differentially rotating disc. A more detailed description includes additional components: bulge, thick disc, stellar halo and dark matter halo, with the latter dominating the overall gravitational potential. Each component is in a different dynamical state, imprinted during the formation history of the Galaxy. After a brief introduction to the main properties of our Galaxy, including basic details of positional astronomy (galactic coordinates and the local standard of rest), the chapter dives into the standard approach of measuring galactic rotation and to the definition of Oort's constants. A perturbation analysis of idealized circular orbits leads to harmonic oscillations, and to the concept of epicyclic motion. In addition to rotation, stars experience vertical motion, perpendicular to the plane of the disc. This motion is also treated in a perturbative way. In the context of the distribution function, one can identify the separation of motion on the plane and perpendicular to it as a fundamental isolating integral (the so-called third integral). The chapter finishes with applications of Jeans equations to the Milky Way and introduces typical functions adopted to describe the potential of the Galaxy.

4.1 General description of the Galaxy

Our Galaxy (see figure 4.1) – commonly written with a capital 'G', to differentiate it from other galaxies – is a late-type system, i.e., a disc galaxy, with a fairly typical luminosity, around $2 \times 10^{10} L_\odot$ (where $L_\odot$ represents the solar luminosity). The chemistry and kinematics of its stellar populations reflect several components, with a different formation history, namely:

- Bulge/Bar: Stars in the central region of the Galaxy have different properties than those in the solar neighbourhood, in terms of the distribution of their ages and chemical composition, as well as their kinematics. In contrast to disc stars, bulge stars have a higher fraction of their kinetic energy in the form of random motion. In addition, there is a bar component, with a significant fraction of radial motion. We know from chapter 3 that stellar orbits keep track of the past dynamical history. Stars in the bulge encode its past formation both through the orbits and the chemical composition, revealing an early and intense formation process (hence the old ages and high metallicity of bulge stars). The stellar mass in the bulge comprises roughly one-fourth of the total.

- Thin disc: This is the component where the solar system is embedded, and it dominates the stellar mass budget of the Galaxy. It can be described by a flat structure, with vertical scalelength ~ 0.3 kpc, and an exponentially decaying profile along the radial direction, with scalelength ~ 2.5 kpc, stretching out to 15–20 kpc. The Sun is located at a distance $R \sim 8$ kpc from the galactic centre. The motion of stars in the thin disc is dominated by bulk rotation, with a roughly constant tangential velocity in the neighbourhood of the Sun, around 220 km s^{-1}, therefore orbiting the Galaxy with a period of $\sim$220 Myr. Stellar populations of different ages feature different velocity dispersions and vertical scalelengths, with the youngest stars having smaller vertical motion and lower dispersion.

- Thick disc: This is the second disc-like component of the Galaxy, with a significantly larger vertical extent (~ 1 kpc), featuring older stars with different chemical composition (higher [Mg/Fe], see section 6.4), although the transition between the thin and the thick disc is very gradual, so it is difficult to identify stars – especially close to the Galactic plane – as belonging to either the thin or the thick disc.

- Stellar halo: Made up of preferentially old and metal-poor stars, moving with high velocities and populating a large spheroidal volume. Its mass contribution to the Galaxy is relatively minor, amounting to about 1 per cent of the total mass in stars.

- Dark matter halo: Although the observations can target only the stellar or gaseous component, we should keep in mind that the whole system is embedded in a much larger structure made up of dark matter particles. In chapter 7 we will explore in more detail the role of dark matter in galaxy formation. We can

Figure 4.1   Gaia/DR2 view of the Milky Way galaxy.
(Source: courtesy European Space Agency/Gaia/Data Processing and
Analysis Consortium.)

simplify this component as a spheroidal or triaxial system that
extends much farther out that any of the above components, to
~150 kpc, comprising about 95 per cent of the total mass content
of the Galaxy. However, note that as it occupies a larger volume,
the dark matter mass density is rather low. Such low densities
explain why, for instance, within the comparatively minute volume
spanned by the solar system, the contribution from dark matter is
negligible.

The coordinate system

The position of stars in the sky is determined by assuming a so-called
celestial sphere; i.e., irrespective of their distance to us, we assign two
angular coordinates, equivalent to the longitude-latitude system that pin-
points a location on Earth. These coordinates are defined in various ways,
depending on the adopted frame of reference. The most important ones
in astrophysics are the following:

1. Alt-Azimuthal coordinates: The simplest choice uses the local
   frame of reference; i.e., the vertical angle above the horizon is the
   elevation, and the horizontal angle – measured from due South – is
   the azimuth. These coordinates are useful to locate a star in the sky

at night, but depend on the time of observation and the location of the observer.

2. Equatorial coordinates: The Earth's rotation determines the frame of reference. The horizon is replaced by the celestial equator, i.e., the projection of the Earth's equator on the celestial sphere. The 'vertical' angle with respect to the equator is the declination. The 'horizontal' angle, measured along the equator from the vernal point – the place where the Sun crosses the celestial equator at the (Northern) Spring equinox[1] – is the right ascension. This is the standard coordinate system for all extragalactic astrophysical sources, as it is independent of the time or location of observation, although it is affected by the long-term precession and nutation of the Earth's rotation axis, requiring the definition of a reference epoch (e.g., J2000.0).

3. Ecliptic coordinates: This system uses the ecliptic as the frame of reference, i.e., the projection on the celestial sphere of the Earth's orbit around the Sun. The ecliptic latitude and longitude are the equivalent angles, where the zero point on the ecliptic is also the vernal point (note the ecliptic plane and the celestial equator intersect at the vernal point, and its opposite position, i.e., the location of the Sun six months later, at the (Northern) Autumn equinox).

4. Galactic coordinates: This is the most convenient way to describe the location of sources in our Galaxy, also given as longitude and latitude. The reference is the projection of the galactic plane on the sky, and the zero point of longitude is the position of the galactic centre.

One can translate the coordinates of a celestial object among these different systems via a set of rotations. In addition to the standard location on the sky, the positions of galactic objects are also typically described by a right-handed Cartesian {X,Y,Z} coordinate system, where the origin is at the solar position, the X axis points towards the Galactic centre, the Y axis points along the direction of rotation and the Z axis is perpendicular to the disc plane.

Exercise 4.1

Where on Earth would the Alt-Azimuthal and equatorial coordinate systems be one and the same?

4.2 Differential rotation in the Galaxy

The motion of stars in the solar neighbourhood and, by extension, in the disc can be described by a differential rotational motion. Most of the kinetic energy in the stars of late-type (i.e., disc) galaxies is in the form of bulk rotation. Note this treatment cannot apply to bulge or halo stars. Historically, this description of motion provided the first steps towards our understanding of the structure of the Galaxy, with astronomers like Oort, Kapteyn and Lindblad pioneering the field. Two different types of stellar motion can be observed. On the celestial sphere we can measure the proper motion with respect to distant, i.e., fixed, sources. This angular motion on the celestial sphere can be mapped onto a physical motion on the normal plane with respect to our line of sight, hence defining a tangential velocity. Along the line of sight it is also possible to measure a radial velocity via the Doppler shift of spectral lines. Both components need to be referred to a fixed frame of reference. The first correction needs to take into account the motion of the Earth around the Sun, which introduces a seasonal variation along the ecliptic of $\Delta v_{ORB} = \pm 30 \, \mathrm{km \, s^{-1}}$. Furthermore, these heliocentric velocities need to take into account the orbital motion of the Sun around the Galactic centre. We therefore need to define a Galactic local standard of rest.

Local standard of rest

In order to understand the motion of stars in the Galaxy, we need to describe their observed velocities as measured from a frame considered "at rest" with respect to the bulk rotational motion of the Galaxy. There are two main definitions of such a local standard of rest (LSR), and it is important to note the difference:

- Kinematic LSR (kLSR): Defined by a frame of reference located at the position of the Sun, moving with the *average* velocity of the stars in the neighbourhood. This is the easiest system from the observational point of view, as it requires only knowledge of the distribution of stellar velocities of nearby stars.
- Dynamical LSR (dLSR): Defined as a frame of reference also located at the solar position, but moving in a circular orbit on the Galactic plane, whose velocity is dictated by the gravitational potential. This definition is the natural one from a theoretical point of view, but we will see that kLSR and dLSR are not the same and that they move apart. This effect is termed 'asymmetric drift', and we will describe it below.

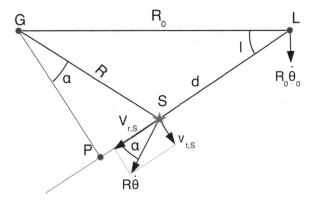

Figure 4.2 Outline of the system involving a nearby star (S), the Sun (L), representing the observer, and the Galactic centre (G).

Lowest order motion: Oort's constants

The simplest model assumes that all disc stars move on the Galactic plane along circular orbits following the gravitational potential. An arbitrary potential is adopted, although for simplicity we will consider spherical symmetry. Note at this point that the potential relates to the total mass content. Therefore, the dark matter halo must be included.

We want to describe the motion of a disc star from our vantage point. The observed motions will be given with respect to the Sun, assumed to follow a simple circular orbit around the Galaxy. Figure 4.2 outlines the measurement, where L is our (heliocentric) reference, S is the target star, and G is the Galactic centre. We denote $\dot{\theta}_0$ and $\dot{\theta}$ the angular velocity of the circular orbits of Sun and star, respectively. We begin by writing the radial and tangential motion of the stars, *relative to L*:

$$\left.\begin{aligned} v_r &= R\dot{\theta}\cos\alpha - R_0\dot{\theta}_0\sin\ell, \\ v_t &= R\dot{\theta}\sin\alpha - R_0\dot{\theta}_0\cos\ell \end{aligned}\right\}. \tag{4.1}$$

Eliminating α from segments $\overline{GP}$ and $\overline{LP}$ and making a Taylor expansion, assuming S is close to L (i.e., $d \ll R_0$), we find, to the lowest order:

$$\left.\begin{aligned} v_r &= -\frac{1}{2}R_0\left(\frac{d\dot{\theta}}{dR}\right)_{R_0}d\sin 2\ell, \\ v_t &= -\frac{1}{2}R_0\left(\frac{d\dot{\theta}}{dR}\right)_{R_0}d\cos 2\ell - \left[\dot{\theta} + \frac{1}{2}R_0\left(\frac{d\dot{\theta}}{dR}\right)_{R_0}\right]d \end{aligned}\right\}. \tag{4.2}$$

These two equations are usually written

$$v_r \equiv Ad \ \sin 2\ell,$$
$$v_t \equiv Ad \ \cos 2\ell + Bd$$

(4.3)

where A and B are Oort's constants of Galactic rotation, defined at $R = R_0$:

$$A = \frac{1}{2} \left[\frac{v_\perp}{R} - \left(\frac{dv_\perp}{dR} \right) \right]_{R_0}$$
$$B = -\frac{1}{2} \left[\frac{v_\perp}{R} + \left(\frac{dv_\perp}{dR} \right) \right]_{R_0}$$

(4.4)

Oort's 'constants' therefore represent a first order description of the rotation curve of the Galaxy. Note that A and B in general are not constants, but depend on the location of the observer – much in the same way as Hubble's constant depends on the cosmological epoch of the observer (see chapter 7).

Exercise 4.2

Show that an isothermal sphere yields $A = -B = 1/_2\Omega_0$; a spherical distribution with homogeneous density results in $A = 0$, $B = -\Omega_0$; and a Keplerian potential gives $A = 3/_4\Omega_0$, $B = -1/_4\Omega_0$, where $\Omega_0 = v_0/R_0$ is the angular speed measured at $R = R_0$.

In the solar neighbourhood, Oort's constants are:[2] $A = 15.3 \pm 0.4 \ \mathrm{km \ s^{-1}}$ $\mathrm{kpc^{-1}}$ and $B = -11.9 \pm 0.4 \ \mathrm{km \ s^{-1} \ kpc^{-1}}$, which, according to the three options in exercise 4.2, suggests a density distribution closer to an isothermal sphere.

Exercise 4.3

Leaving aside potential issues of stability, contrast the values of Oort's constants for a homogeneous spherical distribution and a homogeneous disc (i.e., flat) distribution.

FUNDAMENTALS OF GALAXY DYNAMICS, FORMATION AND EVOLUTION

A more general approach to Oort's constants

The derivation shown above is restricted to the case where the orbits are perfectly circular. Hence, at each point the velocity is purely tangential and constant at fixed radius. We can extend this simple model to a more general case – still assuming motion on the plane – if we write the velocity, measured from the dLSR, as

$$\vec{v}(\vec{R}) = \mathbf{H}(\vec{R}_0) \cdot (\vec{R} - \vec{R}_0) + \mathcal{O}(|\vec{R} - \vec{R}_0|^2). \tag{4.5}$$

Note this is a general Taylor expansion of the velocity field, retaining only the linear terms. The matrix $\mathbf{H}$ can be interpreted as a Jacobian between velocity and position. In Cartesian coordinates,

$$\mathbf{H}(\vec{R}_0) = \begin{pmatrix} \partial_x v_x & \partial_y v_x \\ \partial_x v_y & \partial_y v_y \end{pmatrix}_{\vec{R}_0} \equiv \begin{pmatrix} K+C & A-B \\ A+B & K-C \end{pmatrix}, \tag{4.6}$$

where the second expression defines four Oort's constants, including our previous definitions of A and B, this time extended to the general case, where a radial component of the velocity is present, and where $v_\perp$ can depend on the azimuthal angle. We used the simplified notation $\partial_x \equiv \partial/\partial_x$, and so on. In cylindrical coordinates, the generalized Oort's constants are

$$\left. \begin{aligned} A &= \frac{1}{2}\left[\frac{v_\perp}{R} - \partial_R v_\perp - \frac{1}{R}\partial_\theta v_\parallel \right]_{\vec{R}_0} \\[2mm] B &= \frac{1}{2}\left[-\frac{v_\perp}{R} - \partial_R v_\perp + \frac{1}{R}\partial_\theta v_\parallel \right]_{\vec{R}_0} \\[2mm] C &= \frac{1}{2}\left[-\frac{v_\parallel}{R} + \partial_R v_\parallel - \frac{1}{R}\partial_\theta v_\perp \right]_{\vec{R}_0} \\[2mm] K &= \frac{1}{2}\left[\frac{v_\parallel}{R} + \partial_R v_\parallel + \frac{1}{R}\partial_\theta v_\perp \right]_{\vec{R}_0} \end{aligned} \right\}. \tag{4.7}$$

Note that A and B are equivalent to our previous definition in equation 4.4, when $v_\parallel = 0$ and $\partial_\theta v_\perp = 0$. Furthermore, in the simplified case $C = K = 0$. If we keep the axial symmetry but introduce a non-zero radial term, e.g., radial migration, Oort's constants A and B remain unchanged, but C and K will be non-zero, although potentially small with respect to A and B, as long as this migration term is small compared to the rotation velocity. In the solar neighbourhood, the quoted values of the new constants[3] are: $C = -3.2 \pm 0.4 \, \text{km s}^{-1} \, \text{kpc}^{-1}$; and $K = -3.3 \pm 0.6 \, \text{km s}^{-1} \, \text{kpc}^{-1}$.

Exercise 4.4

Consider a galaxy with a homogeneous density distribution, where stars are assumed to move on circular orbits. Via the Doppler effect, we observe the radial velocity of a star at distance d and Galactic longitude (ℓ). The distance from the observer (O) to the Galactic centre (GC) is R_0. Find the radial velocity as a function of distance (d) from the observer. Do the same for an isothermal sphere distribution.

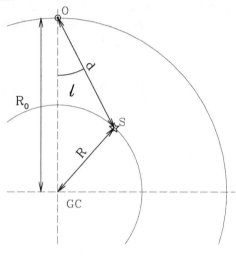

The next order: Epicycles

Let us go back to our original assumption of uniform circular motion (i.e., $C = K = 0$), but now introduce small perturbations about this motion. As is common practice in physics, a first order correction – i.e., retaining only linear terms in the equations of motion – will lead to harmonic oscillations that will allow us to understand the more complex trajectories in the real system. We still constrain the orbits on the Galactic plane, but the radial and tangential components are displaced with respect to the 'equilibrium' case of a circular orbit with radius R_0 (see figure 4.3):

$$\left. \begin{array}{l} \xi \equiv R - R_0 \\ \eta \equiv R_0(\theta - \theta_0) \end{array} \right\}. \qquad (4.8)$$

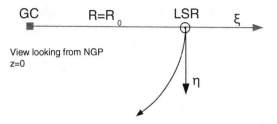

Figure 4.3 Coordinate system used to derive epicyclic motion.

Both perturbed coordinates are assumed to be small with respect to R_0, allowing us to perform a Taylor expansion and retain only the linear terms. The position $(R_0, \theta_0, z = 0)$, also called the 'guiding centre', follows a circular orbit with angular speed

$$\frac{d\theta_0}{dt} = \frac{v_{\perp,0}}{R_0}. \tag{4.9}$$

The acceleration of the motion can be approximated by the centripetal acceleration of the circular orbit:

$$\vec{a} \simeq -\hat{e}_R \frac{v_\perp^2}{R}. \tag{4.10}$$

Now, the acceleration in cylindrical coordinates (see exercise 4.5) is

$$\vec{a} = (\ddot{R} - R\dot{\theta}^2)\hat{e}_R + (2\dot{R}\dot{\theta} + R\ddot{\theta})\hat{e}_\theta + \ddot{z}\hat{e}_z. \tag{4.11}$$

Assuming the dominant component is only the centripetal term leads to

$$\ddot{R} - R\dot{\theta}^2 = -\frac{v_\perp^2}{R}. \tag{4.12}$$

Furthermore, conservation of J_z (the vertical projection of the angular momentum vector) gives

$$R^2\dot{\theta} = R_0 v_{\perp,0}. \tag{4.13}$$

Equation 4.13 can be used to eliminate the $\dot{\theta}$ term in equation 4.12. Finally, we write the equation in terms of the small displacements (equation 4.8) giving

$$\ddot{\xi} = \frac{R_0^2 v_{\perp,0}^2}{R^3} - \frac{v_\perp^2}{R}. \tag{4.14}$$

After some algebra, expanding $v_\perp(R)$ in a Taylor series and keeping only the first order terms, we can write this equation with respect to Oort's constants of rotation:

$$\ddot{\xi} = 4B(A - B)\xi, \tag{4.15}$$

which is the equation of a simple harmonic oscillator, with solution:

$$\xi(t) = H \sin \kappa (t - t_0). \tag{4.16}$$

This motion includes a new frequency, termed the 'epicyclic frequency', namely:

$$\kappa = 2\sqrt{-B(A-B)}. \tag{4.17}$$

The rate of change of ξ is therefore

$$\dot{\xi} = H\kappa \cos \kappa (t - t_0) \rightarrow H = \frac{(v_{\parallel})_{R_0}}{\kappa}. \tag{4.18}$$

To describe the tangential motion (along the η direction), we begin with the conservation of angular momentum (equation 4.13), written in terms of ξ, to lowest order:

$$\dot{\theta} \simeq \frac{v_{\perp,0}}{R_0} \left(1 - \frac{2\xi}{R_0} \right). \tag{4.19}$$

The first term in brackets simply tracks the dLSR, whereas the second term gives the motion relative to the dLSR. Taking only the latter, and writing down the solution to $\xi(t)$ gives:

$$\Delta\dot{\theta} = -\frac{2v_{\perp,0}}{R_0^2 \kappa} \left(v_{\parallel} \right)_{R_0} \sin \kappa (t - t_0). \tag{4.20}$$

The tangential separation of the star and the dLSR is then

$$\eta = \int R_0 \Delta\dot{\theta}\, dt = \frac{2v_{\perp,0}}{\kappa^2 R_0} \left(v_{\parallel} \right)_{R_0} \cos \kappa (t - t_0), \tag{4.21}$$

which can be written in terms of Oort's B constant

$$\eta = -\frac{\left(v_{\parallel} \right)_{R_0}}{2B} \cos \kappa (t - t_0). \tag{4.22}$$

Using equation 4.22 along with equations 4.16 and 4.18, we find the relative motion of nearby stars around the dLSR as a retrograde epicyclic orbit, with axial ratio $\beta \equiv \Delta\eta/\Delta\xi = -\kappa/2B$ (see figure 4.4). The axis ratio can also be written

$$\beta = \sqrt{1 - \frac{A}{B}}. \tag{4.23}$$

The dLSR acts as the guiding centre of the epicycle. Notice that the characteristics of the epicyclic orbits depend purely on the behaviour of the Galactic rotation curve at R_0.

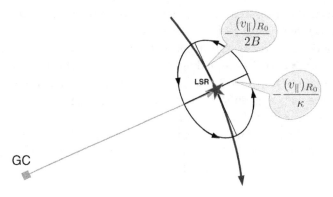

Figure 4.4 Epicyclic motion.

Exercise 4.5

Show that in cylindrical coordinates, the velocity can be written

$$\vec{v} = \dot{R}\hat{e}_R + R\dot{\theta}\hat{e}_\theta + \dot{z}\hat{e}_z,$$

and the acceleration is

$$\vec{a} = (\ddot{R} - R\dot{\theta}^2)\hat{e}_R + 2(\dot{R} - \dot{\theta} + R\ddot{\theta})\hat{e}_\theta + \ddot{z}\hat{e}_z.$$

Using the derived values of A and B, we find the epicycles have a period $\tau_r = 2\pi/\kappa \approx 170\,\text{Myr}$ and axis ratio $\beta = 1.51$ (ellipse elongated along the tangential direction). Given that the 'Galactic year' is 226 Myr, nearby stars take 0.75 galactic rotation periods to complete an epicycle around the dLSR.

Noting that the typical speed along the radial direction is $(v_\parallel)_{R_0} \sim 30\,\text{km/s}$, we find the dimensions of the ellipse are 1.26 kpc along the η direction, and 0.83 kpc along the ξ direction. So, most stars in the solar neighbourhood come from within a radial distance approximately $R_0 \pm 1$ kpc (from the Galactic centre).

Following a star along its epicycle, we note that the ratio of average speeds is

$$\frac{\langle|v_\parallel|\rangle}{\langle|v_\perp - v_{\perp,0}|\rangle} = \frac{\langle d\xi/dt\rangle}{\langle d\eta/dt\rangle} = \frac{(v_\parallel)_0(-2B)\langle\cos\kappa t\rangle}{(v_\parallel)_0\kappa\langle\sin\kappa t\rangle} = \frac{1}{\beta} = 0.68. \qquad (4.24)$$

However, this measurement cannot be derived from the observations. Instead, we compute the average over many stars moving in the solar neighbourhood, each having its own epicentric radius.

Some stars will have guiding centres $R_g > R_0$, i.e., farther away from the position of the dLSR. Those stars will move in the solar vicinity faster along the tangential direction, because of angular momentum conservation. Other stars will have guiding centres $R_g < R_0$, showing up in the solar neighbourhood with slower tangential speeds. The relative tangential speed can be written (using conservation of angular momentum, and keeping only lowest order terms in ξ):

$$\dot{\eta} = R_0(\dot{\theta} - \dot{\theta}_0) \simeq R_0 \left[\dot{\theta}(R_g) + 2\xi \frac{\dot{\theta}(R_g)}{R_g} - \dot{\theta}_0 \right] = \cdots = -2B\xi. \qquad (4.25)$$

Taking an average of the square of the tangential velocity gives

$$\langle \dot{\eta}^2 \rangle = 4B^2 \langle \xi^2 \rangle = \frac{4B^2}{\kappa^2} \langle \dot{\xi}^2 \rangle = -\frac{B}{A-B} \langle \dot{\xi}^2 \rangle. \qquad (4.26)$$

Hence, the ratio of RMS speeds is

$$\frac{\langle v_\perp^2 \rangle}{\langle v_\parallel^2 \rangle} = \frac{-B}{A-B} \simeq 0.46, \qquad (4.27)$$

which is compatible with the observations. Note the epicycles are more elongated along the tangential direction, whereas the velocity distribution is elongated along the radial direction.

If we note that the number density of stars is higher at small radii, we expect more stars with $R_g < R_0$, i.e., moving with slower tangential speeds in the solar neighbourhood: this produces the so-called asymmetric drift. This effect is stronger in older stellar populations (they have higher random speeds, so their orbits deviate further from circular motion).

Exercise 4.6

A galaxy is described by a homogeneous, spherical distribution of stars with total mass M_0 and radius R_0. Find the ratio between the angular velocity and the epicyclic frequency. Can such a potential

---

FUNDAMENTALS OF GALAXY DYNAMICS, FORMATION AND EVOLUTION

produce closed orbits (i.e., orbits that repeat themselves after a number of periods)?

4.3 Vertical motion

We can make a simple estimate of the force normal to the Galactic plane by assuming an infinitely thin, plane-parallel slab with constant density, ρ. Let us assume the vertical motion is independent of the epicyclic motion on the plane (figure 4.5). At point P above the plane, the acceleration due to the volume element $dAdz$ is

$$\vec{da} = \frac{G\rho dz dA}{r^2}\hat{e}_r, \tag{4.28}$$

whose z component is

$$da_z = \vec{da} \cdot \hat{e}_z = -\frac{z}{r}\frac{G\rho dAdz}{r^2} = -G\rho dz d\Omega, \tag{4.29}$$

and we use the solid angle of the volume element as viewed from P:

$$d\Omega = \frac{dA}{r^2}\frac{z}{r}. \tag{4.30}$$

Given that the slab is infinite and homogeneous, we can trivially integrate in solid angle, getting the total vertical acceleration from the whole mass plane as

$$da_z = -2\pi G\rho dz. \tag{4.31}$$

Note that the acceleration is independent of z. The disc of the Galaxy has a vertical extent, so we consider as the next approximation, two thin slabs symmetrically located above and below the $z = 0$ plane, at some distance $\pm z$, with the same density: $\rho(-z) = \rho(z)$.

$$\text{So, for } z(P) > z \qquad da_z = -4\pi G\rho dz,$$
$$\text{whereas for } z(P) < z \qquad da_z = 0, \tag{4.32}$$

and all material farther out from the Galactic plane does not affect the acceleration at P. Hence, the total acceleration can be written as

$$a_z = -4\pi G \int_0^{z=P} \rho(z)dz. \tag{4.33}$$

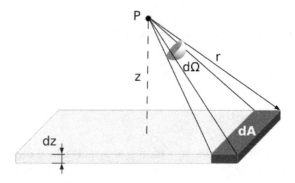

Figure 4.5 Vertical motion.

Near $z = 0$ a constant density profile can be adopted ($\rho \sim \rho_0$). This is a good approximation in the region $|z| \lesssim 100\,\mathrm{pc}$. Hence, at small separations from the Galactic plane, the acceleration will increase linearly with slope $4\pi G\rho_0$. If we define Σ as the surface mass density of the disc, i.e.,

$$\Sigma \equiv \int_{-z_{\mathrm{DISC}}}^{+z_{\mathrm{DISC}}} \rho(z)dz, \tag{4.34}$$

then the acceleration at $|z|$, far from the disc, will be $-2\pi G\Sigma$.

However, the observed vertical acceleration is found to increase because of an extended component (a halo). By examining the density of stars in the halo, it is evident that most of the matter causing this behaviour must be in the form of dark matter. Hence, the density of dark matter can also be derived from vertical motions of stars in the Galaxy. The slope da_z/dz at $z \gtrsim 2\,\mathrm{kpc}$ is especially useful. However, if the observed z is very large, the infinite plane approximation will break down.

Let us now derive the vertical motion for small displacements. If z is small, one can assume $\rho \sim \rho_0$, and

$$a_z = \frac{d^2z}{dt^2} = -4\pi G\rho_0 z, \tag{4.35}$$

describing a simple harmonic oscillator with solution:

$$z = z_{MAX}\,\sin[v(t - t_0)]. \tag{4.36}$$

Differentiating, equation 4.35 gives

$$v^2 = 4\pi G\rho_0 \qquad \text{frequency,}$$

$$\tau_z = \frac{2\pi}{v} = \sqrt{\frac{\pi}{G\rho_0}} \qquad \text{period.} \tag{4.37}$$

For $\rho_0 \sim 0.15 M_\odot pc^{-3}$, the period of the vertical motion is $\tau_z = 68$ Myr. We can now estimate the maximum excursion from the plane, z_{MAX}, for stars in harmonic motion. We just need to evaluate the velocity of the stars when crossing the Galactic plane. At $z \sim 0$, we observe $v_z \sim 6$ km/s. Comparing $z(t)$ and dz/dt we find $v_z^{MAX} = \nu z_{MAX}$, giving $z_{MAX} \sim 67$ pc, reached after $\tau_z \sim 17$ Myr.

Therefore, stars in our neighbourhood are expected to probe a toroidal region centred on the LSR, with radial extent $\Delta R = \pm 250$ pc over a timescale $\tau_r = 175$ Myr; and a vertical extent $\Delta z = \pm 67$ pc, with timescale $\tau_z = 68$ Myr. If any of the periods involved in this motion coincide, e.g., $\tau_r(R) = 1/2$ of a Galactic year, then resonance ensues, leading to density anomalies such as spiral density waves (see chapter 5).

4.4 The collisionless Boltzmann equation in galactic coordinates

In order to apply the CBE to the Milky Way, we take advantage of the axisymmetric shape, writing the equation in cylindrical coordinates (R, ϕ, z), where the distribution function reads

$$f(\vec{r}, \vec{v}; t) = f(R, \phi, z, v_\parallel, v_\perp, v_z; t), \qquad (4.38)$$

and the CBE can be written

$$\partial_t f + \dot{R}\partial_R f + \dot{\phi}\partial_\phi f + \dot{z}\partial_z f + \dot{v}_\parallel \partial_{v_\parallel} f + \dot{v}_\perp \partial_{v_\perp} f + \dot{v}_z \partial_{v_z} f = 0. \qquad (4.39)$$

The general equations of motion for a test particle in cylindrical coordinates – note ϕ (azimuthal angle) $\neq \Phi$ (potential) – are

$$\left.\begin{array}{l} \dot{v}_\parallel = -\dfrac{\partial \Phi}{\partial R} + \dfrac{v_\perp^2}{R} \\[2ex] \dot{v}_\perp = -\dfrac{1}{R}\dfrac{\partial \Phi}{\partial \phi} - \dfrac{v_\parallel v_\perp}{R} \\[2ex] \dot{v}_z = -\dfrac{\partial \Phi}{\partial z} \end{array}\right\}, \qquad (4.40)$$

leading to

$$\partial_t f + v_\parallel \partial_R f + \frac{v_\perp}{R}\partial_\phi f + v_z \partial_z f + \left(\frac{v_\perp^2}{R} - \partial_R \Phi\right)\partial_{v_\parallel} f -$$

$$-\frac{1}{R}\left(v_\parallel v_\perp + \partial_\phi \Phi\right)\partial_{v_\perp} f - \partial_z \Phi \partial_{v_z} f = 0. \qquad (4.41)$$

We make the following assumptions: (1) the system is in steady state ($\partial_t f = 0$); (2) there is full axial symmetry: $\partial_\phi = 0$ (for any function). Now the CBE is:

$$v_\| \partial_R f + v_z \partial_z f - \left(\partial_R \Phi - \frac{v_\perp^2}{R}\right) \partial_{v_\|} f - \frac{v_\| v_\perp}{R} \partial_{v_\perp} f - \partial_z \Phi \partial_{v_z} f = 0. \quad (4.42)$$

This CBE (an equation in partial derivatives involving five variables) can be broken into four first-order differential (subsidiary) equations:

$$\frac{dR}{v_\|} = \frac{dz}{v_z} = \frac{dv_\|}{\left(\frac{v_\perp^2}{R} - \partial_R \Phi\right)} = \frac{dv_\perp}{\left(-\frac{v_\| v_\perp}{R}\right)} = \frac{dv_z}{\left(-\partial_z \Phi\right)}. \quad (4.43)$$

Therefore, a maximum of four isolating integrals can be invoked. Two are obvious from the symmetries of time (steady state) and rotation about the z axis, namely, energy and the projection along $\hat{e}_z$ of angular momentum:

$$\mathcal{I}_1 = E = \frac{1}{2}\left(v_\|^2 + v_\perp^2 + v_z^2\right) + \Phi(R, z), \quad (4.44)$$

$$\mathcal{I}_2 = J_z = R v_\perp. \quad (4.45)$$

If these two are the only isolating integrals, Jeans theorem implies that

$$f(R, z, v_\|, v_\perp, v_z) = F(E, J_z). \quad (4.46)$$

Notice $v_\|$ and v_z enter only into $\mathcal{I}_1$ and only as a term $(v_\|^2 + v_z^2)$; hence they should be interchangeable in the analysis, leading to the same velocity dispersion:

$$\langle v_\|^2 \rangle = \langle v_z^2 \rangle. \quad (4.47)$$

The observations of the motion of stars in the Milky Way lead to a different outcome:

$$0.4 \langle v_\|^2 \rangle = \langle v_z^2 \rangle, \quad (4.48)$$

suggesting an additional isolating integral in the system.

Vertical motion and the third isolating integral

The description of stellar trajectories in the galaxy suggests it is possible to separate between motion on the disc and motion perpendicular to it, in the flattened potential of a disc. This behaviour reveals the presence of a third isolating integral. Close to the Galactic plane, the z component of the force – which depends on the local mass density – is almost decoupled from the R component – which depends on the enclosed mass. We have seen this behaviour in sections 4.2 and 4.3, where the epicycles and the vertical motion of stars moving close to the Galactic plane can be treated as independent harmonic oscillators. The potential is, therefore, expected to be separable into an R-dependent function and a z-dependent function, namely:

$$\Phi(R, z) \simeq \Phi_1(R) + \Phi_2(z), \tag{4.49}$$

which allows us to solve one of the subsidiary equations of the CBE (see equation 4.43):

$$\frac{dz}{v_z} = \frac{dv_z}{-d\Phi_2/dz}, \tag{4.50}$$

resulting in an integral of motion:

$$\mathcal{I}_3 = \frac{1}{2}v_z^2 + \Phi_2(z), \tag{4.51}$$

which can be defined as the energy involving vertical motion (being independently conserved with respect to the energy related to rotation). If we take this as a third isolating integral, the distribution function for a flattened axisymmetric distribution (such as our Galaxy) now reads

$$f(R, z, v_\parallel, v_\perp, v_z) = F(E, J_z, \mathcal{I}_3), \tag{4.52}$$

whose dependence on v_z is no longer the same as on $v_\parallel$, breaking the symmetry between $\langle v_\parallel^2 \rangle$ and $\langle v_z^2 \rangle$.

4.5 Application of Jeans equations

There are several applications of Jeans equations to the Milky Way. In chapter 3 we presented the derivation of Jeans equations as a marginalization of the CBE in velocity space. The continuity equation (see equation 3.54) can be written in cylindrical coordinates as follows:

$$\frac{\partial \nu}{\partial t} + \frac{1}{R}\frac{\partial (R\nu\langle v_\parallel \rangle)}{\partial R} + \frac{\partial (\nu\langle v_z \rangle)}{\partial z} = 0. \tag{4.53}$$

The derivation of the set of Jeans equations involving the first order moment of the velocities (from the general expression in equation 3.58) is more involved but rather straightforward, and leads to the following:

$$\left.\begin{array}{l} \dfrac{\partial(\nu\langle v_\parallel\rangle)}{\partial t} + \dfrac{\partial(\nu\langle v_\parallel^2\rangle)}{\partial R} + \dfrac{\partial(\nu\langle v_\parallel v_z\rangle)}{\partial z} + \nu\left(\dfrac{\langle v_\parallel^2\rangle - \langle v_\perp^2\rangle}{R} + \dfrac{\partial\Phi}{\partial R}\right) = 0 \\[3mm] \dfrac{\partial(\nu\langle v_\perp\rangle)}{\partial t} + \dfrac{\partial(\nu\langle v_\perp v_\parallel\rangle)}{\partial R} + \dfrac{\partial(\nu\langle v_\perp v_z\rangle)}{\partial z} + \dfrac{2\nu}{R}\langle v_\parallel v_\perp\rangle = 0 \\[3mm] \dfrac{\partial(\nu\langle v_z\rangle)}{\partial t} + \dfrac{\partial(\nu\langle v_z v_\parallel\rangle)}{\partial R} + \dfrac{\partial(\nu\langle v_z^2\rangle)}{\partial z} + \nu\left(\dfrac{\langle v_\parallel v_z\rangle}{R} + \dfrac{\partial\Phi}{\partial z}\right) = 0 \end{array}\right\}.$$

$$(4.54)$$

We will show below two applications of Jeans equations to the dynamics of the Milky Way.

Vertical motion and Jeans equation

The first example concerns the equation corresponding to the vertical component of velocity. Let us assume stationary state – i.e., neglecting the partial derivative in time – and decorrelation between motion on the plane and vertical motion – leading to $\langle v_z v_\parallel\rangle = 0$. The latter approximation is especially relevant for motion close to the Galactic plane and is well justified by our definition of the third integral. The last equation in 4.54 now reads

$$\frac{1}{\nu}\frac{\partial(\nu\langle v_z^2\rangle)}{\partial z} = -\frac{\partial\Phi}{\partial z} = \frac{d^2 z}{dt^2} = -4\pi G\int_0^z \rho(s)\,ds, \qquad (4.55)$$

where we have used our results from section 4.3 about the vertical acceleration in the infinite plane-parallel slab approximation. In this expression, the density and RMS of the vertical velocity can be observed, allowing us to determine the local mass density. Oort obtained a value of the density $\rho(R_0, z = 0) \sim 0.15\,M_\odot\,\mathrm{pc}^{-3}$ (Oort limit). Notice the density requires a double differentiation, whereas the surface mass density needs only one derivative (i.e., it is less uncertain). Oort's estimate was $\Sigma(R_0, |z| < 0.7\,\mathrm{kpc}) \sim 90\,M_\odot\,\mathrm{pc}^{-2}$.

Asymmetric drift

The kLSR and dLSR (see section 4.2) are expected to be one and the same if all stars move along circular orbits dictated by the gravitational potential. However, their positions slowly drift with time if we adopt epicyclic motion instead. We can give an adequate representation of this effect. Let us consider the radial velocity component of Jeans equation (the first equation in 4.54), with the additional assumption of symmetry of the density profile along the vertical direction. For positions close to the plane, we have $\partial_z \nu \sim 0$, and so

$$\frac{R}{\nu}\frac{\partial(\nu\langle v_\parallel^2\rangle)}{\partial R} + R\frac{\partial\langle v_\parallel v_z\rangle}{\partial z} + \langle v_\parallel^2\rangle - \langle v_\perp^2\rangle + R\frac{\partial\Phi}{\partial R} = 0. \qquad (4.56)$$

We define the azimuthal velocity dispersion as

$$\sigma_\phi^2 \equiv \langle(v_\perp - \langle v_\perp\rangle)^2\rangle. \qquad (4.57)$$

After some algebra, and noting that the velocity of the dLSR is $v_c^2 = R(\partial\Phi)/\partial R$, we arrive at the asymmetric drift equation:

$$v_c^2 - \langle v_\perp\rangle^2 = \langle v_\parallel^2\rangle\left[-\frac{\partial\ln(\nu\langle v_\parallel^2\rangle)}{\partial\ln R} - \left(1 - \frac{\sigma_\phi^2}{\langle v_\parallel^2\rangle}\right) - \frac{R}{\langle v_\parallel^2\rangle}\frac{\partial(\langle v_\parallel v_z\rangle)}{\partial z}\right].$$
$$(4.58)$$

Note that v_c tracks the motion of the dLSR, whereas $\langle v_\perp\rangle$ corresponds to the average circular velocity for a number of stars (i.e., the kLSR). Let us define $v_{\mathrm{kLSR}} \equiv \langle v_\perp\rangle$ and $v_{\mathrm{dLSR}} \equiv v_c$. Also, since $|v_{\mathrm{dSR}} - v_{\mathrm{kLSR}}| \ll v_{\mathrm{dLSR}}$, we can write

$$v_{\mathrm{dLSR}}^2 - v_{\mathrm{kLSR}}^2 \sim 2v_{\mathrm{dLSR}}(v_{\mathrm{dLSR}} - v_{\mathrm{kLSR}}), \qquad (4.59)$$

and so

$$v_{\mathrm{dLSR}} - v_{\mathrm{kLSR}} \sim \frac{\langle v_\parallel^2\rangle}{2v_{\mathrm{dLSR}}}\cdot[\cdots]. \qquad (4.60)$$

The terms within the brackets are the same ones as in equation 4.58. The first term in brackets is the dominant one. The last term can be neglected for our purposes. As the density decreases outwards, the first term in brackets is positive, i.e., $v_{\mathrm{dLSR}} > v_{\mathrm{kLSR}}$, and so, the kLSR lags behind the dLSR. Empirically, $v_{\mathrm{dLSR}} - v_{\mathrm{kLSR}} \sim \langle v_\parallel^2\rangle/D$, with $D = 120\ \mathrm{km\,s}^{-1}$.

4.6 The potential of the Galaxy

The observed rotation velocity profile of disc galaxies (see figure 4.6) can be described to the lowest order by three regimes, as sketched in the right panel of the figure. In the central region, we have a linearly rising portion, which is the standard rigid body motion expected when the mass density is constant (see section 2.5). Most of the rotation curve of the galaxy is relatively flat, suggesting an isothermal density profile. The sketch also shows the expected Keplerian decrease of the rotation velocity in the regions where the mass density drops to zero. However,

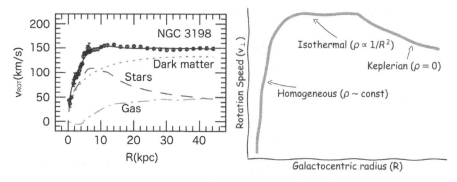

Figure 4.6 Comparison of the observed rotation velocity profile of a disc galaxy (*left*, NGC3198), with a simple model involving three regions (*right*). The data for NGC3198 also show the individual contribution of gas, stars and dark matter to the rotation curve. (Source: adapted from Dutton et al., 2005, ApJ, 619, 218.)

observations of dynamical tracers in the outer regions of galaxies (using, e.g., halo stars, planetary nebula or globular clusters) do not seem to reach such regions, suggesting that the dark matter halo is much more extended than any of these tracers. The observational data in the figure show the individual contribution of gas and stars to the rotation curve. Note that if we write the circular velocity as $v_\perp^2 = GM(<R)/R$, we can add the individual contributions to the orbital velocity in quadrature, namely: $v_\perp^2 = v_{stars}^2 + v_{gas}^2 + v_{DM}^2$. Aside from such simple models to describe the galaxy, one can create more complex mass distributions to mimic the rotation curve (and any other dynamical information, such as Oort's constants) as accurately as possible. The Schmidt model represents one of the early attempts,[4] with a linear superposition of nonhomogeneous spheroidal mass distributions, along with a point mass, leading to a rotation profile:

$$v_\perp^2(R) = \frac{GM_P}{R} + 4\pi G \sum_i \sqrt{1 - e_i^2} \int_0^R \frac{\sum_i \rho_i(a)a^2}{\sqrt{R^2 - a^2 e_i^2}} da, \qquad (4.61)$$

where each component is defined by a spheroid with eccentricity e_i and density $\rho_i(a)$, and a is the semi-major axis of each spheroid. In fact, this model built upon a previous one defined by Oort that assumed only constant density spheroids.[5] More complex models are currently adopted to describe our Galaxy. As an example, we show below the potential-density

FUNDAMENTALS OF GALAXY DYNAMICS, FORMATION AND EVOLUTION

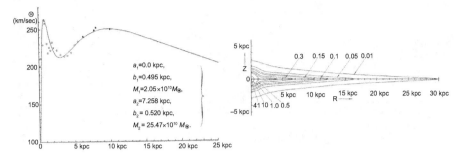

Figure 4.7 Miyamoto-Nagai model of the Galaxy. The left panel shows the best-fit rotation curve that consists of two components, with parameters a, b and M shown in equation 4.62: one corresponding to the bulge (subindex 1) and another one representing the galactic disc (subindex 2). The panel on the right shows a vertical cut of the density distribution, with contours in units of $M_\odot\,\mathrm{pc}^{-3}$. (Source: Miyamoto & Nagai, 1985, PASJ, 27, 533.)

pair corresponding to the Miyamoto-Nagai model:[6]

$$
\left.\begin{aligned}
\Phi(R,z) &= -\frac{GM}{\sqrt{R^2 + \left(a^2 + \sqrt{b^2 + z^2}\right)^2}} \\
\rho(R,z) &= \frac{b^2 M}{4\pi}\,\frac{aR^2 + [a + 3\sqrt{z^2 + b^2}][a + \sqrt{z^2 + b^2}]^2}{\{R^2 + [a + \sqrt{z^2 + b^2}]^2\}^{5/2}(z^2 + b^2)^{3/2}}
\end{aligned}\right\}. \qquad (4.62)
$$

Note the limit $a \to 0$ represents a spherical Plummer model, and $b \to 0$ describes a very flattened potential. For a specific combination of two models (i.e. two choices of a, b and the associated mass), one can obtain a good fit to the rotation curve of our Galaxy (see figure 4.7).

Notes

1 I.e., when the Sun moves from the Southern to the Northern celestial hemisphere.
2 Bovy, 2017, MNRAS, 468, 63.
3 Bovy, 2017, MNRAS, 468, 63.
4 Schmidt, 1956, Bulletin of the Astronomical Institutes of the Netherlands, 13, 15.
5 Oort, 1952, ApJ, 116, 233.
6 Miyamoto & Nagai, 1985, PASJ, 27, 533.

5
Specific aspects of disc and elliptical galaxies

This chapter presents a few properties of the dynamics of stars in the two general morphological classes of galaxies: disc and elliptical systems. Their different dynamical states (discs supported by rotation; ellipticals supported mostly by random motion) and surface brightness radial gradients (with discs having exponential distributions and ellipticals featuring steeper profiles) are telltale signatures of their different formation histories. We begin with a general overview of their main properties. We explore the scaling relations in galaxies (Tully-Fisher in discs, Fundamental Plane and its projections in ellipticals), and their connection with the virial theorem, as well as other physical processes. Regarding ellipticals, we will briefly present the concept of random motion support and its connection with the observed ellipticities of galaxies. A brief section is included on spiral structure in disc galaxies, revisiting the epicyclic motion of stars on the disc, and presenting a scenario (density wave theory), where orbital resonance can lead to the rigid body rotation of a pattern (the spiral structure).

5.1 'Hot' versus 'Cold' dynamical systems

We saw in chapter 1 that galaxies can be morphologically classified into three main families: spheroidal, disc-like and irregular. The morphological appearance of a galaxy gives a direct link to its dynamical state. Detailed studies of the kinematics of the gaseous and stellar components reveal a substantial difference. If we split the total kinetic energy of the stars in a galaxy between ordered motion (rotation) and random motion (velocity dispersion), we find that spheroidal galaxies are supported mostly by velocity dispersion, whereas disc galaxies keep most of their kinetic energy in the form of bulk rotation. If we use a simple analogy between velocity dispersion and temperature, one can say that this classification divides the sample into 'hot' and 'cold' dynamical systems, or slow and fast rotators, respectively. There are two parameters often used to quantify this split: the ratio (v/σ), typically measured within an effective radius (i.e., the region within which half of the total flux is observed), and the specific angular momentum parameter:

$$\lambda \equiv \frac{\langle R|v|\rangle}{\langle R\sqrt{v^2+\sigma^2}\rangle}. \tag{5.1}$$

Incidentally, the traditional morphological separation between ellipticals and spirals as presented by the Hubble tuning fork diagram does not convey the true dynamical evolution of galaxies, and an alternative classification scheme was proposed to take into account angular momentum in addition to morphology.[1] Note, for instance, that only one-third of nearby ellipticals shows a hot dynamical state.

Late-type galaxies

Late-type galaxies (LTGs) consist of the family of disc galaxies, featuring spiral arms of any type (termed Sa, Sb, Sc, Sd in the classical notation) and barred systems (SBa, SBb, SBc, SBd). The subclass label from a to d refers to a gradation from close, tightly wound spiral arms (a) to more open arms (d). Lenticular galaxies (S0), although classified as disc systems, are not considered late-type galaxies. In addition to the disc component, late-types feature a central bulge. The so-called classic bulges have dynamical and population properties similar to those of ellipticals. However, a second type of bulge (pseudo-bulge) is expected to form instead, after the onset of dynamical instabilities within the galaxy. The bulge-to-disc ratio – which can be measured either by mass or luminosity – decreases along the Hubble sequence (largest for S0, smallest for Sd galaxies).

If we invoke conservation of angular momentum, the collapse and subsequent cooling of the baryonic material is expected to produce a galaxy with a significant amount of rotation (i.e., a disc). In addition, processes that feed star formation through the streaming of low-entropy gas along filaments (cold accretion) will also introduce rotation. Rotating structures resembling discs are found one way or the other over a wide range of cosmic distances, although disc galaxies at high redshift appear dynamically hotter than disc galaxies observed locally.[2]

Early-type galaxies

Elliptical galaxies (E) and lenticulars (S0) constitute the other fundamental group, termed 'early-type galaxies' (ETGs). These systems are
mostly hot dynamical systems (even lenticular discs feature prominent,
dynamically hot bulges). We will see below that their spheroidal morphology is supported by orbital anisotropy. Ellipticals are characterized
by their apparent flattening (although note that the shape is dependent
on projection), ranging from E0 to E7, where the number corresponds to
$10 \times (1 - b/a)$ with b being the projected short axis and a the projected
long axis. Overall, their stellar content is dominated by a roughly homogeneously old and metal-rich population, suggesting an early and rapid
process of star formation. This result led to the belief that massive elliptical galaxies could form from a monolithic collapse of gas.[3] Lower mass
ETGs have younger, metal-poor populations and may even feature faint
levels of residual star formation – following the general trend of downsiz-
ing, where the bulk of star formation progresses from massive to low-mass
galaxies with cosmic time.

The standard view of the formation of elliptical galaxies invokes a
merging process where the progenitors can be either ellipticals or disc
galaxies,[4] but should have comparable mass (major mergers). Additional
growth mechanisms can be invoked via minor mergers, i.e., the infall of a
low-mass satellite, but these processes are not expected to trigger a morphological change (although they can heat up the distribution of orbits,
as in, e.g., the thick disc of the Milky Way). This merging scenario for
the formation of ellipticals is consistent with the morphology-density relation,[5] whereby elliptical galaxies are preferentially found in high density
environments. The structure growth in these environments evolves at a
faster rate in comparison with the field; therefore, the prevalence of ellipticals reflects the role of interactions among structures as a way to explain
their formation. Compact massive galaxies are already found at redshifts
$z \sim 2$, i.e., when the Universe had roughly one-fourth of its present age,
and are viewed as the progenitors of the cores of massive ellipticals and
bulges at present time. However, the details of the formation process, including the role of major and minor mergers, and the initial cold accretion
phase are poorly understood.

5.2 Scaling relations

The observable properties of galaxies have strong correlations that indic-
ate, to a lowest order, a similar formation process driven by a reduced

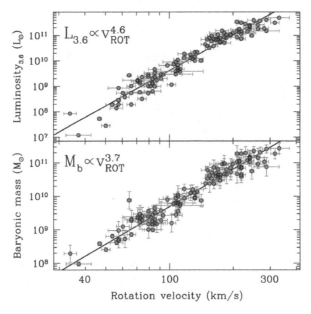

Figure 5.1 Tully-Fisher relation from the SPARC disc galaxy sample. (Source: adapted from Lelli, McGaugh & Schombert, 2016, ApJ, 816, L14.)

number of parameters, most notably the mass of the galaxy (either total or stellar) or the average velocity dispersion of the stars. A blind search for correlations among galaxy properties via principal component analysis found that many of the properties can be determined from a single parameter![6]

The Tully-Fisher relation

One of the most important scaling relations in disc galaxies is the correlation between total luminosity and rotation velocity (figure 5.1).[7] The velocity that defines this relation is usually taken in the flat portion of the rotation curve, away from the centre. The scatter of this correlation is small enough to make it a standard candle and has been used to measure cosmological distances. The Tully-Fisher relation (TFR) can be determined from a simple dynamical argument invoking the virial theorem:

$$v_{ROT}^2 \propto \frac{M}{R} \Rightarrow L \propto \Upsilon^{-2}\Sigma^{-1}v_{ROT}^4. \tag{5.2}$$

However, this result implies that the TFR requires constant mass-to-light ratio across all disc galaxies, as well as constant surface brightness. The

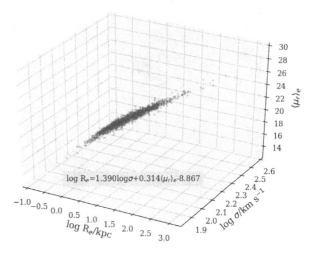

Figure 5.2 Fundamental plane of a sample of early-type galaxies compiled from the Sloan Digital Sky Survey. (Source: adapted from La Barbera et al., 2010, MNRAS, 408, 1335.) Used by permission of Oxford University Press.

latter is found to apply to the subset of high surface brightness galaxies (Freeman's law). The relation depends on wavelength, increasing the power index of v_{ROT} towards the redder bands. Notice that at longer (i.e., redder) wavelengths, the light from the stellar component is mostly contributed by low-mass (mainly older) stars. Hence, variations in Υ across the sample are much smaller than at bluer wavelengths. The correlation with respect to luminosity features a break at low-luminosities, with galaxies falling below the relation defined for the more massive galaxies (i.e., lower luminosity at fixed v_{ROT}). The correlation is extended to lower mass galaxies when using total baryonic (stellar plus gas) mass rather than luminosity, showing that the trend could, in principle, be fully related to the baryonic mass; this result has been used as an argument to propose alternative models of gravitation that require no dark matter.[8]

The Fundamental Plane and its projections

The Fundamental Plane (FP) is a three-parameter correlation involving the size, velocity dispersion and surface brightness of early-type galaxies. These estimates are typically averaged inside an aperture extending over the effective radius or a fraction thereof. The data span a plane on this three-dimensional parameter space (see figure 5.2). By use of the virial theorem, it is possible to define a plane that appears to be tilted with respect to the observed FP. This mismatch can be related to several factors.

(1) A change in the mass-to-light ratio, $\Upsilon \propto M^{\alpha}$ (see exercise 5.1), can explain this tilt. Such an increase in M/L can be driven by a gradual increase of the contribution from dark matter towards more massive galaxies. (2) Rotation will "divert" part of the kinetic energy into bulk rotation, so that a systematic change in v/σ with galaxy mass could introduce a similar effect. (3) The simplified model presented in exercise 5.1 also assumes that all galaxies feature the same surface brightness profile (i.e., their light distribution is homologous). A systematic trend, for instance, a change of the Sérsic index with galaxy mass, will produce a tilt of the plane defined by the simple model based on the virial theorem. The actual reason for the tilt of the FP is arguably a combination of the three.[9]

Exercise 5.1

The Fundamental Plane is an important scaling relation of elliptical galaxies, where the average surface brightness (Σ), the size (R) and the velocity dispersion (σ) are related by the equation

$$R \propto \sigma^{1.2} \Sigma^{-0.8}$$

(these numbers have been slightly changed to make the problem consistent as is). Using the mass-to-light ratio $\Upsilon = M/L$ to convert light into mass, show that the virial theorem leads to $R \propto \sigma^2 \Sigma^{-1} \Upsilon^{-1}$.

Let us now assume that the mass to light ratio changes systematically from galaxy to galaxy, such that $\Upsilon \propto M^{\alpha}$. Find the value of α that brings the virial expectation in line with the observations.

In addition to the three-parameter relation spanned by the Fundamental Plane, projections defined by just two parameters also provide interesting scaling relations of early-type galaxies, although not as tight, by definition, as the FP.

1. Faber-Jackson relation: This trend involves the total luminosity (i.e., a combination of surface brightness and size) and the velocity dispersion:[10]

$$L \propto \sigma^{\alpha}, \tag{5.3}$$

and a typical value of the power law index is $\alpha \sim 4$. It is the equivalent of the Tully-Fisher relation in disc galaxies. Naively, one could guess this relation by assuming that while the dominant contribution to the kinetic energy of discs comes from rotation, the kinetic energy in early-type galaxies appears mostly in the form of velocity dispersion. However, there is a significant variation of the degree of rotational support in ETGs that correlates with mass, so this trend is not so trivial.

2. Kormendy relation: This is the 'easiest' scaling relation from the observational point of view,[11] as it compares size (R) and surface brightness (Σ). Therefore it does not require any spectroscopic data. It can also be expressed by a simple formula:

$$\Sigma \propto R^{\gamma}, \qquad\qquad (5.4)$$

and γ is approximately -3, i.e., the surface brightness decreases with galaxy size. This scaling relation deeply encodes information about the dynamical history of ETGs. In simple words, the Kormendy relation states that bigger galaxies (broadly more massive galaxies) are 'fluffier'.

Exercise 5.2

Using exercise 3.7, explain how a scenario that postulates the formation of early-type galaxies only via mergers can explain the Kormendy relation.

The colour-magnitude relation

In addition to the above relations, which depend mainly on galaxy dynamics, there is a scaling relation pertaining to the properties of the stellar populations. The colour-magnitude relation is especially significant in early-type galaxies, mostly lying on the red sequence (see figure 1.5). The more massive galaxies have redder colours. This relation has been exploited, for instance, to detect galaxy clusters, as they host a large fraction of elliptical galaxies. The redness of the colour of a population can be explained either by its (older) age, (higher) metallicity or (dustier) interstellar medium. The latter can be ruled out as elliptical galaxies do not have much gas (or dust) in them. A combination of the first two factors is thought to explain this trend, so that the star formation history of early-type galaxies is closely related to its total mass (more accurately its velocity dispersion). A detailed analysis based on spectroscopic observations from the Sloan Digital Sky Survey revealed a strong correlation between the age[12] and the metallicity[13] of a galaxy with respect to its mass. This overly simple trend gets more complicated when dealing with colour gradients. In addition to morphology and the global dynamical state of a galaxy, radial gradients provide further information about the assembly of galaxies, but this is something beyond the scope of this book.

5.3 Rotation versus 'pressure' in early-type galaxies

The tensor virial theorem, presented in section 3.9, allows us to relate the contribution from rotation and random motion to the shape of the galaxy, which, in turn, affects the gravitational potential energy. If the system is in equilibrium, there is no change in the moment of inertia tensor, and the virial theorem can be written

$$2T_{jk} + \Pi_{jk} + W_{jk} = 0, \tag{5.5}$$

where the kinetic energy is split between a streaming motion component (T) and a random motion component (Π). Using Cartesian coordinates and adopting the z direction as the axis of rotation, we can write

$$T_{xx} = T_{yy},$$
$$T_{ij} = 0 \qquad \text{if } i \neq j, \tag{5.6}$$

and likewise for Π and W. Hence, we only have two nontrivial equations for 5.5:

$$\left.\begin{array}{l} 2T_{xx} + \Pi_{xx} + W_{xx} = 0 \\ 2T_{zz} + \Pi_{zz} + W_{zz} = 0 \end{array}\right\}. \tag{5.7}$$

We can therefore relate the kinetic and potential energy terms as follows:

$$\frac{2T_{xx} + \Pi_{xx}}{2T_{zz} + \Pi_{zz}} = \left|\frac{W_{xx}}{W_{zz}}\right| \sim \left(\frac{a}{b}\right)^{0.89}, \tag{5.8}$$

where the final expression in this equation is derived by taking into account a spheroidal distribution of matter.[14] Let us consider the two possible options that can explain the flattening of an elliptical galaxy:

1. Rotation: If rotation dominates the kinetic energy budget and fully explains the flattening of ellipticals, we can assume the velocity dispersion tensor is isotropic:

$$\Pi_{xx} = \Pi_{yy} = \Pi_{zz} = M\sigma_0^2, \tag{5.9}$$

where σ_0 is the mass-weighted velocity dispersion of the random motion, along the line of sight to the galaxy. The streaming component of the kinetic energy is

$$T_{xx} + T_{yy} = \frac{1}{2} \int \rho \langle v_\phi^2 \rangle d^3x = \frac{1}{2} M v_0^2, \tag{5.10}$$

and v_0^2 is the mass-weighted mean square rotation speed. Equation 5.8 gives

$$\frac{v_0}{\sigma_0} = \sqrt{2 \left(\frac{a}{b} \right)^{0.89} - 2}. \tag{5.11}$$

2. **Anisotropic velocity dispersion:** The flattening can also be caused by a difference in the distribution of orbits (dynamical systems remember their past history, in contrast to a system in thermodynamic equilibrium). In this scenario, the streaming term vanishes:

$$T_{xx} = T_{yy} = T_{zz} = 0, \tag{5.12}$$

and we divide the random motion between a contribution on the rotation plane, and perpendicular to it:

$$\left. \begin{aligned} \Pi_{xx} &= \Pi_{yy} = M\sigma_x^2 \\ \Pi_{zz} &= M\sigma_z^2 \end{aligned} \right\}, \tag{5.13}$$

giving

$$\frac{\sigma_z}{\sigma_x} = \left(\frac{b}{a} \right)^{0.45}. \tag{5.14}$$

In the extreme case of an E7 elliptical, $1 - b/a = 0.7$, a full rotational support would require $v/\sigma \sim 1.96$, i.e., a rather large rotation velocity that is not observed in ellipticals, whereas a mild anisotropy ($\sigma_z/\sigma_x \sim 0.58$) can explain the same shape. Figure 5.3 shows the relation between v/σ and ellipticity in a sample of nearby early-type galaxies, where the solid dots represent slow rotators (i.e., hot dynamical systems) and the open symbols are fast rotators. The dashed line traces our simple expression from equation 5.11. A more accurate expression of the rotational support limit is shown as a solid line, roughly enveloping all the observed data points.[15]

5.4 A brief introduction to spiral arms in disc galaxies

Disc galaxies often feature spiral arms. In fact, one of the main criteria for the classification of galaxies in the Hubble tuning fork diagram (see figure 1.3) is the morphology of these spiral features. From the observational point of view, the nature of spiral structure is correlated

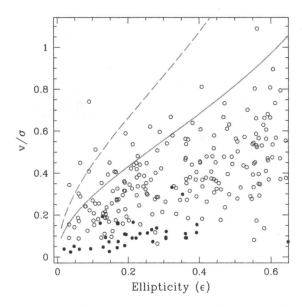

Figure 5.3 Rotational support of early type galaxies as a function of ellipticity. (Source: data from the ATLAS^{3D} sample, Emsellem et al., 2011, MNRAS, 444, 888.) Used by permission of Oxford University Press.

with the global properties, such as the ratio of gas to stellar mass. Gas poor discs (lenticular galaxies: S0) have no spiral structure. Most spirals feature two arms with conspicuous dust lanes on their inner edge. There are two main types of spiral morphology: grand design, with arms that can be traced over a large radial range, and flocculent, showing patchy, discontinuous spiral features (figure 5.4). The star formation rate is significantly higher in the spiral arms than in the rest of the galaxy. A consequence of this is the prominent display along the spiral arms of hot (OB, i.e., very young $< 10\,\mathrm{Myr}$) stars, as well as ionized gas (HII regions).

The winding-up paradox

A first guess at explaining spiral features would resort to differential rotation in the disc. We saw in chapter 4 that a significant part of the rotation curve of our Galaxy is flat. Hence the orbital speed is independent of radius, and the angular speed is $\Omega(R) \propto 1/R$. This will imply that an extended burst of star formation would gradually twist into a spiral feature as the central regions will wind faster than the outer ones. Consider a stripe of stars, all at the same azimuthal angle $\phi_0 = 0$ at time $t = 0$. At a later time, the angular distribution will be $\phi(R, t) = \Omega(R)t$. The differential

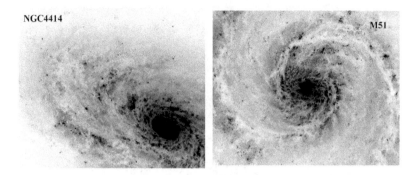

Figure 5.4 Spiral arm morphology: flocculent (*left*) versus grand design (*right*). (Source: courtesy NASA/ESA HST and the Hubble Heritage Project.)

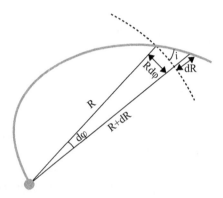

Figure 5.5 Determining the pitch angle of a spiral feature. The dashed line follows a circumference, for which the pitch angle is $i = 0$.

rotation will therefore shear the straight line into a spiral with a pitch angle (i), defined as the angle between the tangent to the spiral and the circle crossing at that point (figure 5.5):

$$\cot i = \left| R \frac{d\phi}{dR} \right| = Rt \left| \frac{d\Omega}{dR} \right| = 2At, \qquad (5.15)$$

where we have made use of Oort's A constant (equation 4.4). If the pitch angle is small, we can relate it to the separation, ΔR, between arms. Take two adjacent arms at the same azimuthal angle:

$$2\pi = |\Omega(R + \Delta R) - \Omega(R)|t \Rightarrow \Delta R = \frac{2\pi R}{\cot i} = \frac{2\pi}{t} \left| \frac{d\Omega}{dR} \right|^{-1} = \frac{\pi R}{At}, \quad (5.16)$$

where the last step assumes $\Delta R \ll R$. The separation decreases monoton-ically with time. Plugging in expected values for the Milky Way (t$\sim$10 Gyr; $At = 151.36$, $R = 8.5$ kpc), we find $\Delta R \sim 0.2$ kpc, which is in conflict with the observational constraints. This result corresponds to a very small pitch angle ($i \sim 0.2°$), whereas the observed value is around 5° for the earlier-type systems (Sa), increasing to $10° - 30°$ for the later-type ones (Sc, Sd). Hence, either the spiral arms are not long lived or the spiral pattern is not directly linked to differential rotation.

Spiral pattern rotation

A possible explanation for the presence of long-lived spiral arms invokes a resonance mechanism, such that stars are preferentially found near the spiral pattern, but, on average, the composition of stars within a spiral arm changes with time. An analogous example, closer to home, is the pile up of cars on a congested motorway. As cars approach the traffic jam, they will slow down, increasing the local density of cars. On the other side of the congestion, cars will accelerate, decreasing the density as they move away. If we look from above, we will see an overdensity of cars in a seg-ment of the motorway, but different cars join and leave this overdensity with time. Is it possible to produce a similar pattern in galaxies? From chapter 4, we know that stars in a disc galaxy do not move along circular orbits. They follow instead an elliptical orbit (an epicycle) that trails along a guiding centre representing the simple circular orbit (i.e., the location of the dLSR). The epicyclic motion has frequency $\kappa = 2\sqrt{-B(A - B)}$, in con-trast with the angular frequency of the dLSR (Ω). Let us assume there is a pattern of stars in a galaxy, i.e., an inhomogeneous distribution of stars (say a set of spiral arms), which moves with a constant pattern speed Ω_P. If the number $m \equiv (\Omega - \Omega_P)/\kappa$ is a rational number, the trajectories will be closed and may allow for this density pattern to be sustained. The choice $m = 2$ is especially relevant, as it creates a two-armed structure. For a given choice of the pattern angular speed, there are three import-ant radial positions (figure 5.6, right panel). (1) At the position where $\Omega_P = \Omega$ (called corotation, CR), stars move with the pattern. Therefore, from a frame of reference rotating with angular speed Ω_P, we will see the stars simply tracing an epicycle. (2) At $\Omega_P = \Omega - \kappa/2$ (Inner Lindblad Res-onance, ILR), a star will move twice round its epicycle for every rotation around the galaxy, overtaking the pattern. (3) At $\Omega_P = \Omega + \kappa/2$ (Outer Lindblad Resonance, OLR), the star moves twice its epicycle for every rotation, falling behing the pattern. At any other radii, the orbits are un-closed trajectories. The left panel of figure 5.6 shows the radial variation

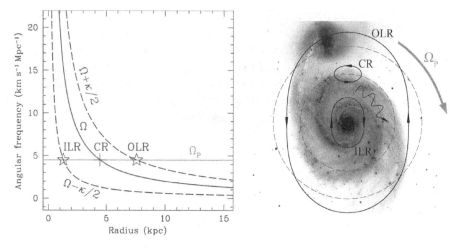

Figure 5.6 Illustration of Lindblad resonances. (Source: background image of M51 from the Digital Sky Survey.)

of angular speed for an isothermal density profile; other profiles will look similar. Notice the $\Omega - \kappa/2$ curve is flat outside of R $\sim$ 4 kpc. If the pattern speed is similar to this value, the galaxy will extend its ILR over a large radial range; i.e., it would be possible to have oval-like orbits at many radii – when viewed from a frame of reference moving with the pattern speed.

Figure 5.7 shows two different cases where an extended ILR can produce a pattern in solid body rotation. The figure nests orbits similar to the ones described above for the ILR over a range of radii. We expect these orbits to rotate with constant angular speed Ω_P. In the first case (left panel) the orientation is constant from the inside out, producing a bar-like feature that rotates rigidly. The second case (right panel) introduces a gradual rotation of the orientation of the orbits, producing a clear spiral feature, also expected to rotate rigidly. This explanation is purely kinematic, based on the motion of stars. It is meant only to motivate the case for a resonance that can cause spiral features. Density wave theory details the mechanism by which the interaction of gas and stars in a disc galaxy can give rise to spiral arms in solid body rotation.[16] The source of energy to drive the wave is still a matter of debate, and tidal interactions with a nearby object or the presence of a rotating bar may be involved. Note, for instance, that galaxy M51, shown in the right panel of figure 5.6, is tidally affected by a nearby galaxy (to the top of the figure), and many other grand-design spirals also have nearby galaxies that could be potential interlopers. We have considered in the example above a

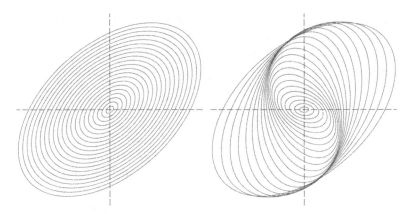

Figure 5.7 Rigidly rotating features expected when the ILR is extended. Either a bar (*left*) or a spiral (*right*) is produced, depending on the radial variation of the orientation of the orbits.

two-armed spiral by choosing $m = 2$, but in general we could consider a pattern with $m > 2$. Stars and gas clouds will pass through the pattern with frequency $m[\Omega_p - \Omega(R)]$ perturbing the system. These motions with respect to the spiral pattern will affect the gravitational potential of the disc. It is found that stars respond so as to strengthen the spiral only if the perturbing frequency is slower than κ. Hence, a spiral density wave can propagate only in the region where this condition holds, i.e.,

$$\Omega - \kappa/m \leq \Omega_p \leq \Omega + \kappa/m. \tag{5.17}$$

The extrema of this interval are the inner and outer Lindblad resonances. Patterns with many arms ($m > 2$) will therefore span a narrow region in the diagram (e.g., the interval between $\Omega - \kappa/4$ and $\Omega + \kappa/4$ will shrink in figure 5.6), impeding the formation of extended spiral structure with many arms. The vast majority of grand-design spiral galaxies display two arms. Note that at the corotation radius, stars move along with the pattern, whereas inside (outside) corotation they will move faster (slower) than the pattern. When gas clouds hit the spiral pattern, they get compressed, triggering star formation. This explains why most of the young stars are found along the spiral arms. Once they form, the stars overtake the spiral feature (inside corotation). Hence, we should expect an offset between the gas, ongoing star formation, and more evolved stars, following the (orthogonal) distance to a spiral arm. This is one of the characteristic properties exploited to test density wave theory.[17] Gas is more affected by the spiral pattern than stars, because of its smaller random motion. As stars move in the galaxy, they increase their random motion. The distribution of older stars in the disc is more homogeneous (i.e., less affected by the presence of spiral arms). Therefore, lenticular galaxies (S0, disc systems with very little gas) have no spiral structure, whereas later-type systems (Sb, Sc) normally feature the most prominent arms.

Nevertheless, at present, numerical simulations cannot create a spiral pattern with solid body rotation, as predicted by density wave theory. Recent research suggests that spiral arms are short-lived, in corotation with the stars.[18] This scenario would be compatible with the presence of flocculent arms, traditionally explained by the onset of detonation waves caused by star formation, propagating around the disc and getting sheared by differential rotation. However, grand-design spirals pose a challenge to numerical models when tidal interactions are not invoked.

Notes

1 Cappellari et al., 2011, MNRAS, 416, 1680.
2 Förster-Schreiber et al., 2009, ApJ, 706, 1364.
3 Eggen, Lynden-Bell & Sandage, 1962, ApJ, 136, 748, although note that this classic paper focused on the formation of our Galaxy.
4 Toomre & Toomre, 1972, ApJ, 178, 623.
5 Dressler, 1980, ApJ, 236, 351.
6 Disney et al., 2008, Nature, 455, 1082.
7 Tully & Fisher, 1977, A&A, 54, 661.
8 Sanders & McGaugh, 2002, ARA&A, 40, 263.
9 Trujillo, Burkert & Bell, 2004, ApJ, 600, L39.
10 Faber & Jackson, 1976, ApJ, 204, 668.
11 Kormendy, 1977, ApJ, 218, 333.
12 Gallazzi et al., 2005, MNRAS, 362, 41.
13 Tremonti et al., 2004, ApJ, 613, 898.
14 Binney & Tremaine, 2008, *Galactic dynamics*, Princeton, section 2.5.
15 Binney, 1978, MNRAS, 183, 501.
16 Lin & Shu, 1964, ApJ, 140, 646.
17 Tamburro et al., 2008, AJ, 136, 2872.
18 Grand et al., 2012, MNRAS, 426, 167.

6
Galactic chemical enrichment

Most of the chemical elements in the periodic table were synthesized in the cores of stars. The Big Bang, although very hot during a brief period of time, could not keep the required condition of temperature and density for long enough, resulting only in a 25 per cent production by mass of helium from hydrogen, along with traces of a few heavier nuclei. The synthesis in the Universe of elements such as C, O, Mg and Fe can take place only in the central regions of stars, where the conditions for thermonuclear reactions are adequate. Therefore, the build-up of the elements in galaxies – observed in the stellar and gaseous components – reflects the past star formation histories inside galaxies. Tracing the composition of gas and stars provides a powerful tool for understanding galaxy formation. This chapter presents an overview of chemical enrichment along with the differential equations that govern the variation of elemental abundances. The simplification of the instantaneous recycling approximation allows us to produce very simple solutions that give insight about the different scenarios of galaxy evolution. We relate these solutions to the observed scaling relation between mass and metallicity and the solution of the G-dwarf problem in Milky Way stars. Abundance ratios are also introduced as a powerful discriminant of star formation histories.

6.1 Nucleosynthesis and the formation of galaxies

The formation and evolution of galaxies can be split into two complementary channels. One deals with the growth of the dark matter halos within which galaxies live. This growth – the mass assembly history (MAH) – is driven mainly by gravitational interactions, from the growth of the small density fluctuations at early times (chapter 7) to merging processes that gradually create more massive structures. The second channel involves the star formation history (SFH) of the galaxies inhabiting these halos and can be studied via galactic chemical enrichment. As star formation proceeds, the chemical elements synthesized in stars are progressively incorporated into subsequent stellar generations, although a fraction of these elements, along with the H+He gas, can be expelled from galaxies. A simple description of these proceses allows us to understand the mechanisms underlying the transformation of gas into stars. This is arguably one of the most complicated problems in astrophysics, as a large number of physical processes play an important role, including the hydrodynamics of gas inflows and outflows, the cooling and fragmentation of gas clouds, feedback from star formation and activity from an Active Galactic Nucleus (AGN).

Central to the analysis of galactic chemical enrichment (hereafter GCE) is the concept of stellar yields, defined as the mass of a given chemical element synthesized in a star of mass m. The yields are calculated from models of stellar structure and evolution, aspects that are beyond the scope of this textbook. However, the most fundamental aspect is the significantly different role played by massive stars ($\gtrsim 8 M_\odot$). They contribute the majority of chemical elements, and feature comparatively short lifetimes (25 Myr for a $10 M_\odot$ star, in contrast with 10 Gyr for a $1 M_\odot$ star). Therefore, chemical enrichment closely traces the star formation history. The presence of an additional nucleosynthetic channel involving a binary system – type Ia supernovae – introduces an additional timescale that will be presented at the end of this chapter. The specific details of a star formation region can, in principle, leave an imprint on the detailed chemical composition of the stars formed within it. The so-called chemical tagging analysis is a promising method by which GCE could help identify the dynamical origin of a population of stars in our Galaxy (or nearby resolved systems).

6.2 General aspects of galactic chemical enrichment

In the simplest form, the equations of GCE trace the evolution of the gas and stellar mass, as well as the amount of chemical elements. Often, we add together all chemical elements – except for H and He – defining metallicity (Z) as the mass fraction in these elements with respect to the total. As a reference, our Sun has a metallicity $Z_\odot \sim 0.02$. An alternative estimator of metallicity uses an individual element, typically iron, with [Fe/H] defined as the logarithmic ratio, by number, of iron atoms, with respect to the solar value. In general, one can find stars and stellar populations with very low metallicity ([Fe/H] $\lesssim-3$, and even lower), but high metallicity populations are not found above a factor of $\sim 2-3$ times the solar value.

The initial mass function

Of all parameters describing a star, its mass is the most important one (Russel-Vogt theorem). Given a stellar mass, we can infer its lifetime, yields, evolutionary phases and remnant properties. Additional factors such as chemical composition and rotation will also affect these, but we assume that, to a good approximation, only mass defines lifetime and yields, essential parts of the GCE equations. Therefore, a fundamental aspect of chemical enrichment is the distribution of stellar masses in a population at birth. This is the initial mass function (IMF). Local observations of stellar populations support the idea that this function may be universal, although recent evidence in star-forming and quiescent galaxies reveals potential variations, possibly towards a top-heavy IMF in star-forming systems (i.e., an excess of massive stars with respect to the standard IMF); and a bottom-heavy IMF in massive quiescent galaxies (i.e., an excess of low-mass stars with respect to the standard). Nevertheless, we will assume here a unique definition of the IMF, given by a single power law:

$$\phi(m) \equiv \frac{dN}{dm} \propto m^{-\Gamma}, \tag{6.1}$$

where $\Gamma = 2.35$ defines the Salpeter IMF. We also need to consider the mass range of the IMF, i.e., the interval of mass within which stars can be found. The low-mass end is caused by the threshold in the onset of thermonuclear reactions in the stellar core, at around $0.08 M_\odot$. Lower masses do not have enough gravitational energy to achieve the high central temperatures needed. Instabilities in the stellar atmospheres control the upper limit, although the actual value is not clearly defined. A value of $100 M_\odot$ is commonly adopted for this limit. Quite often, the IMF is defined on a logarithmic scale in mass:

$$\xi(m) \equiv \frac{dN}{d \log m} \propto m^{-\mu}, \tag{6.2}$$

with the Salpeter[1] IMF corresponding to a slope $\mu = \Gamma - 1 = 1.35$. We also need to use the normalization mass scale for a given IMF, defined as

$$m_N \equiv \int_{m_{\text{low}}}^{m_{\text{high}}} m\phi(m)dm. \tag{6.3}$$

This normalization is typically fixed at a solar mass, $m_N = M_\odot$, and some textbooks redefine the normalization, removing the mass units from this expression by choosing $m_N = 1$. Note that in this case, the IMF is defined per unit mass, instead of per number of stars. There are a number of additional functional definitions of the IMF, where the main effect is either to taper off the low-mass end with either an additional power law (or sets of power laws), as in the Scalo or Kroupa definitions, or to replace the power law at low masses with a lognormal distribution, as in the Chabrier IMF.

Exercise 6.1

Show that for a Salpeter IMF, defined in the mass interval 0.1 $-100 \, M_\odot$ a supernova is expected per $184 \, M_\odot$ of stars formed (assuming that only stars with mass $m > 10 M_\odot$ undergo a supernova explosion). The star formation rate of our Milky Way galaxy is approximately one supernova per century, and the average star formation rate is $1 \, M_\odot \, \mathrm{yr}^{-1}$. Which way should we change the IMF to reconcile these estimates?

Remnant mass and returned fraction

Once a star reaches its endpoint (after a time τ_m), we assume that a remnant is left behind: a white dwarf for low- and intermediate-mass stars ($m \lesssim 8$–$10 M_\odot$) and a neutron star or black hole for massive stars. This can be quantified by defining a remnant mass (w_m) that remains locked away at later times. The rest of the mass is ejected back into the interstellar medium (ISM), including both 'gas' and 'metals', which become available for the next episodes of star formation. Therefore, if we consider the formation of a single population of stars (all created at the same time), we can define the fraction in mass returned back to the ISM. This so-called returned fraction can be related to the IMF as follows:

$$R(t) \equiv \frac{1}{m_N} \int_{m_t}^{m_{\text{high}}} \phi(m)(m - w_m)dm, \qquad (6.4)$$

where m_t is the stellar mass corresponding to a lifetime t. This function is zero at $t = 0$, gradually increasing with time towards an asymptotic value at late times, when stellar lifetimes are as long as the age of the Universe. Typical values of the returned fraction at late times are ~ 0.3 for the Salpeter IMF and $\sim$0.5 for the Chabrier IMF. Since the latter features a lower fraction of low-mass stars – which effectively lock up most of their mass forever – the amount of gas available for recycling is significantly higher.

Exercise 6.2

Compute the returned fraction at late times of the Salpeter IMF (adopt 1 $M_\odot$ as the low-mass threshold of the integral), if the remnant mass of a star with initial mass m is given by (in solar mass units):

$$\frac{w_m}{M_\odot} = \begin{cases} 0.1(m/M_\odot) + 0.5, & m \leq 10 M_\odot, \\ 1.5, & m > 10 M_\odot. \end{cases}$$

Stellar yields

Another important component of GCE equations is the yield (p_m), defined as the mass fraction in a given element, that is synthesized by the star (of mass m) and returned to the interstellar medium once it reaches its endpoint, after a time τ_m. An equivalent version of the returned fraction is the net yield, defined as

$$y_p \equiv \frac{1}{1-R} \int_{m_t}^{m_{high}} m p_m \phi(m) dm. \tag{6.5}$$

Star formation rate

The star formation rate (ψ) is an essential function in GCE equations, as it drives the transformation from gas into stars. The Schmidt law assumes that ψ is some power law of the gas mass available for star formation. A simple argument gives

$$\psi(t) \propto \frac{\rho_g(t)}{t_{dyn}} \implies \psi(t) \equiv k_{SF} \rho_g^{1.5}(t). \tag{6.6}$$

From an observational point of view, only projected surface mass densities can be measured. The equivalent version of the above on two-dimensional surface formation rates and surface gas mass densities is the Kennicutt law:

$$\Sigma_\psi(t) = k'_{SF} \Sigma_g^{1.4}(t). \tag{6.7}$$

Exercise 6.3

Assuming $L \propto M^{3.5}$ – consider only the main sequence and neglect evolved phases of stellar evolution – compare the average mass and average luminosity of a population at birth and after 1 Gyr (the lifetime of a 2.5 $M_\odot$ star is 1 Gyr). Neglect the contribution to the mass from remnants.

6.3 Basic equations of galactic chemical enrichment

The basic equations of GCE[2] describe the evolution of the mass in stars (M_s) and gas (M_g), as well as the amount of "metals" in the gas phase, described by the metallicity (Z). We assume the total mass is $M = M_s + M_g$. More advanced equations separate the gas component into several 'phases', mainly a cold phase (which controls the star formation rate) and a hot phase. Furthermore, the metallicity can be split into the contribution from individual elements, or from groups closely related to their synthesis reactions, such as the α elements, iron-peak elements, or s- and r-process elements. Gas flows are simplified by two time-dependent functions, an infall rate, $f(t)$ and an outflow rate $o(t)$. The basic equations describing the evolution of the mass components are

$$\left.\begin{aligned} \frac{dM}{dt} &= f - o \\ \frac{dM_s}{dt} &= \psi - E \\ \frac{dM_g}{dt} &= -\psi + E + f - o \end{aligned}\right\}. \tag{6.8}$$

The quantity $E(t)$ is the ejection rate, i.e., the amount of gas contributed by stars at the end of their lives. Noting that w_m is the remnant mass, we can write

$$E(t) = \int_{m_t}^{m_{max}} (m - w_m)\hat{\psi}(t - \tau_m)\phi(m)dm, \tag{6.9}$$

where $\hat{\psi} = \psi/m_N$ is the star formation rate per unit mass, so defined as to give $E(t)$ the units of a mass per unit time. The evolution of the metal content is given by

$$\frac{d(ZM_g)}{dt} = -Z\psi + E_Z + Z_f f - Zo. \qquad (6.10)$$

The enrichment process is described by a homogeneous mixture of gas and metals, with instantaneous mixing. The ejection of metals is given by

$$E_Z(t) = \int_{m_t}^{m_{max}} [(m - w_m)Z(t - \tau_m) + mp_m]\, \hat{\psi}(t - \tau_m)\phi(m)dm, \qquad (6.11)$$

where p_m are the stellar yields, i.e., the mass fraction of a star of mass m converted into metals, and ejected at the end of the stellar life.

The instantaneous recycling approximation

These equations can be simplified if we assume that stars are divided into two classes: those that live forever (masses below some threshold $m < m_0$) and those that die out as soon as they are born ($m > m_0$). This is called the instantaneous recycling approximation (IRA). The approximation is well justified because most of the yields are produced by $M \gtrsim 8M_\odot$ stars, which have much shorter lifetimes than $\sim M_\odot$ stars, which lock most of the mass into stars. The ejecta (see equations 6.9 and 6.11) are now

$$E(t) = R\psi(t),$$
$$E_Z(t) = RZ(t)\psi(t) + y(1 - R)\,[1 - Z(t)]\,\psi(t). \qquad (6.12)$$

The factor $[1 - Z(t)]$ in the second equation is ~ 1, given that the metallicity is always $Z \ll 1$. The equations of chemical enrichment in the IRA become

$$\left.\begin{array}{l} \dfrac{dM_s}{dt} = (1 - R)\psi \\[2mm] \dfrac{dM_g}{dt} = -(1 - R)\psi + f - o \\[2mm] \dfrac{d(ZM_g)}{dt} = -Z(1 - R)\psi + y(1 - R)\psi + Z_f f - Zo \end{array}\right\}. \qquad (6.13)$$

Instead of tracing the mass in metals (ZM_g), one can write the equation for the evolution of the metallicity (Z), as follows:

$$M_g \frac{dZ}{dt} = y(1 - R)\psi + (Z_f - Z)f. \qquad (6.14)$$

Neglecting gas infall and outflow, we see that the total amount of metals ever formed at some time t is

$$Z_s M_s + Z_g M_g = \int_0^t \left[\int_{m_s}^{m_{max}} m p_m \hat{\psi}(s - \tau_m) \phi(m) dm \right] ds = y M_s. \quad (6.15)$$

The closed box model

The above equations can be applied to basic models of chemical evolution to derive analytic expressions of the stellar metallicity distribution. The simplest option is the closed box model, which has no infall or outflows, with initial conditions $M_g(t=0) = M_0$; $Z(t=0) = 0$. The total mass remains constant $M = M_s + M_g = M_0$ at all times. Integrating the IRA equations gives

$$\frac{dZ}{y} = -\frac{dM_g}{M_g} \implies Z = y \ln \mu_g^{-1}, \quad \text{where } \mu_g \equiv \frac{M_g}{M_g + M_s} = \frac{M_g}{M_0}. \quad (6.16)$$

Note this expression is valid as long as $Z \ll 1$, so it will break down at late times, as $\mu_g \to 0$. The closed box model produces a significant amount of low-metallicity stars, which is as expected since the star formation rate at early times (i.e., with low-metallicity gas) is very high. Comparisons of this model with observed stellar metallicities in the solar neighbourhood show a large mismatch, termed the 'G-dwarf problem'. We can quantify this difference by estimating the mass fraction in stars with metallicity below some threshold Z. We need to rewrite the IRA equations, relating the metallicity to the stellar mass:

$$\frac{dZ}{y} = \frac{dM_s}{M_0 - M_s} \implies M_s(<Z) = M_0 \left(1 - e^{-Z/y} \right). \quad (6.17)$$

In order to derive this expression, we use the fact that the metallicity in the gas phase increases monotonically in this model, so that the stellar mass when the gas phase metallicity is Z comprises only stars with metallicities below this value. We can quantify, for instance, the mass fraction in stars with metallicity below one quarter of the solar value. Firstly, we need to derive the yield, by use of the observational constraint of the gas mass fraction in the solar neighbourhood, at present time:

$$\mu_g(\text{NOW}) = \frac{\Sigma_{gas}}{\Sigma_{gas+stars}} \sim \frac{5 M_\odot \, pc^{-2}}{50 M_\odot \, pc^{-2}} = 0.1, \quad (6.18)$$

and the metallicity in the gas phase is at present $Z \sim Z_\odot$. Therefore, from equation 6.16 we find $y = 0.43Z_\odot$, which, when plugged into equation 6.17, gives a mass fraction $M_s(<Z_\odot/4) \sim 0.45$. The observational constraint is ~ 0.02, i.e., a substantially smaller fraction. Therefore, the closed-box model does not give a valid representation of the solar neighbourhood as it overproduces low-metallicity stars.

The open box models: Infall and outflows

To mitigate the G-dwarf problem posed by the closed box model, we need a mechanism to reduce the production of low-metallicity stars. The simplest solution beyond the closed box model is to consider that stars are being formed from the infall of gas supplied by an external reservoir. By assuming a small initial gas mass, we prevent the early locking of a high-mass content in stars when the metallicity is too low. To simplify the model, let us assume no outflows, and infall perfectly balancing the star formation rate, such that the gas mass is kept constant: $M_g = M_0 \Longrightarrow f = (1 - R)\psi$, leading to

$$M_s(<Z) = M_0 \ln \left(\frac{y}{y - Z} \right), \qquad (6.19)$$

where we also assume that the gas reservoir has zero metallicity. As in the closed box case, the net yield can be derived from

$$\frac{dZ}{y - Z} = \frac{dM_s}{M_0} \Longrightarrow \frac{Z}{y} = 1 - e^{1 - \frac{1}{\mu_g}}. \qquad (6.20)$$

Therefore, in the solar neighbourhood ($\mu_g = 0.1$) we get $y \simeq Z_\odot$, and $M_s(<Z_\odot/4) = 0.032$, in line with the observational constraints (see figure 6.1).

Exercise 6.4

Show that the closed box and infall models presented above lead to the following distribution of stars with respect to metallicity, as shown in figure 6.1:

$$\frac{dN}{d\log \zeta} \propto \begin{cases} \zeta e^{-\zeta}, & \text{closed box,} \\ \dfrac{\zeta}{1 - \zeta}, & \text{infall,} \end{cases}$$

where $\zeta \equiv Z/y$.

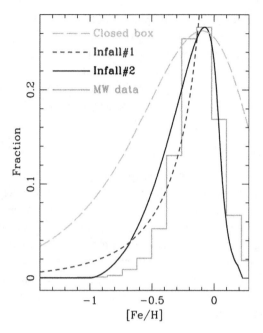

Figure 6.1 Comparison of the observed metallicity of nearby stars from the Geneva-Copenhagen Survey[a], shown in grey, with the closed box and infall model presented here (the latter labelled 'Infall #1'). A second, more realistic distribution from a numerical model of chemical enrichment[b] is shown as 'Infall #2'. Used by permission of Oxford University Press.

---

[a]Holmberg et al., 2007, A& A, 475, 519.
[b]Ferreras, Wyse & Silk, 2003, MNRAS, 345, 1381.

It is also interesting to incorporate the effect of outflows in the chemical enrichment process. For instance, let us assume that the outflow rate is proportional to the star formation rate, defining $o \equiv c\dot{M}_s$, where c is the proportionality constant. This ansatz is justified by the fact that supernovae-driven winds, triggered effectively by the instantaneous star formation rate (SFR) when IRA is adopted, drive the gas outflow rate. If there is no infalling gas, and the initial condition is $M_g = M_0$, and zero stellar mass, we get

$$Z = \frac{y}{1+c} \ln \mu_g^{-1} \equiv y_{eff} \ln \mu_g^{-1}, \qquad (6.21)$$

equivalent to the closed box model, where the effective yield is the net yield, reduced by a factor $(1+c)$. Likewise, the cumulative stellar mass below some metallicity threshold is

$$M_s(<Z) = \frac{M_0}{1+c}\left[1 - e^{-\frac{Z}{y_{\text{eff}}}}\right],$$
(6.22)

again, equivalent to the closed box model. By increasing c, one could bring this mass fraction in line with the observations.

Exercise 6.5

Consider now the leaky box model, with the same properties and initial conditions as the closed box model, but with additional infall (f) and outflow (o) terms, where $f = o = c\dot{M}_s$ (i.e., the infall and outflow rates are balanced and correspond to a fraction, $c < 1$, of the rate of change in stellar mass). Derive the mass fraction in low-metallicity ($Z < Z_\odot/4$) stars, and contrast with the observed ~ 0.02 fraction.

The outflow model can also be presented in the light of the observed mass-metallicity relation (figure 6.2), where the most massive galaxies feature the highest metallicities. If we make a simple relation between the observed gas-phase metallicity and the effective yield, a functional dependence would be established between c – i.e., the gas outflow efficiency – and galaxy mass. Such a relation could be motivated by an escape velocity argument, where the gravitational potential well would 'modulate' the amount of gas ejected in outflows.

Exercise 6.6

A galaxy fuels its star formation via infall, but without any outflowing material. At time $t = 0$ the gas mass is M_0 and the stellar mass is zero. The infall rate is at all times a fixed fraction (α) of the star formation rate ($f = \alpha\psi$), and the star formation rate follows a simple law: $\psi = kM_g$. Find the time evolution of the gas fraction $\mu \equiv M_g/(M_g + M_s)$. What is the asymptotic behaviour of μ as $t \to \infty$? What happens if $\alpha = 1$?

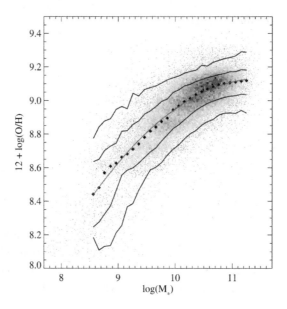

Figure 6.2 Observed correlation between the stellar mass of the galaxy and the (gas-phase) metallicity, measured with respect to the oxygen abundance. In these units, solar metallicity corresponds to a value of 8.7. (Source: adapted from Tremonti et al., 2003, ApJ, 613, 898.)

6.4 Chemistry as a cosmic clock

So far, the GCE equations adopt a single variable to describe metallicity (Z). The next step would entail separating the production of elements into α elements and Fe-peak elements. The main reason for this choice is that the former are mainly produced by core-collapse (Type II) supernovae, which are the final evolutionary stage of massive ($\gtrsim 8\,M_\odot$) stars. In these stars, most of the iron-rich core is trapped into a remnant (neutron star or black hole), locking this material away. The Fe yield of Type II supernovae is generally rather low ($\lesssim 0.1\,M_\odot$ per supernova). In contrast, Type Ia supernovae – triggered in close binary systems involving at least one white dwarf – release a significantly higher amount of Fe ($\sim 0.7\,M_\odot$). Since such a scenario requires the presence of a white dwarf, a delay is expected in the production of Type Ia supernovae with respect to Type II, and thus iron is incorporated later in subsequent generations, as long as star formation is present. Although estimating this delay is a complicated task that depends on a large number of factors involving the production and evolution of binaries and the accretion of material in a way that would yield

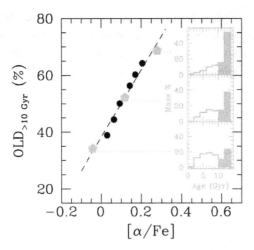

Figure 6.3 Correlation between the fraction of old stars (vertical axis, given as the mass fraction in stars older than 10 Gyr) and the chemical abundance ratio [α/Fe] (where 0 corresponds to the solar value). Note that higher ratios are associated with galaxies where the old component is more prominent, suggesting a relation between the star formation history and chemical enrichment. (Source: adapted from de la Rosa et al., 2011, MNRAS, 418, L74.) Used by permission of Oxford University Press.

a supernova explosion, we can approximate this delay by a Δt_{Ia}, which allows us to use the abundance ratio [α/Fe] as a cosmic clock: galaxies with high stellar [α/Fe] result from an intense, short-lived episode of star formation that shut off quickly, before Type Ia supernovae had any chance to contribute to chemical enrichment. If [α/Fe] is low, the star formation rate was more extended. For most models of the Type Ia progenitor, the delay time (Δt_{Ia}) ranges between 0.5 and 2 Gyr. For example, massive early-type galaxies feature metal-rich and old stellar populations, with substantially higher [α/Fe] than stars in the solar neighbourhood, reflecting a very different star formation history (figure 6.3). The old and high metal content suggests an early and efficient formation process to quickly build up the gas-phase metallicity, and to avoid locking too many low-mass (i.e., long-lived) stars at low metallicity. The high [α/Fe] reinforces the need for a high formation efficiency, requiring most of the stars to be formed within $\lesssim$1–2 Gyr. Massive galaxies are observed at high red-shift, potentially the cores of high-mass early-type galaxies at present. They have a stellar mass of about $10^{11} M_\odot$ at redshift z~2. These systems must sustain a very intense formation scenario, where the average star formation rate needs to be about two orders of magnitude higher than in the Milky Way at present. In the Milky Way there is also a signature of nonsolar abundance ratios, as low-metallicity stars (formed in the early stages of evolution) have high $[\alpha/Fe]$, a property used, for instance, to disentangle the populations of thin and thick disc stars (see section 4.1).

Notes

1 Salpeter, 1955, ApJ, 121, 161.
2 Galactic chemical enrichment was a field pioneered by Beatrice Tinsley (Tinsley, 1980, Fundam. Cosmic Phys., 5, 287).

7
The growth of density fluctuations

Galaxies start as minute density fluctuations in an otherwise homogenous and expanding Universe. During the first phases of galaxy growth, the fluctuations were so small that simple equations can be solved to trace this early evolution (linear phase). An overview of cosmology is necessary, including the treatment of the expanding Universe through Friedmann's equations, although a simple Newtonian argument will suffice here, in lieu of a more complex treatment with general relativity (outside the scope of this textbook). The concept of critical density and the evolution of the background matter and radiation densities will be presented, leading to the equations of an expanding fluid with a small perturbation, parameterized with the density contrast. The Einstein–de Sitter case will be presented as an easy-to-solve model that gives an accurate representation of the Universe during the main phase of linear growth. A comparison with the density fluctuations at present in the Cosmic Microwave Background leads to one of the most robust proofs that dark matter must be present in large quantities in the Universe. The linear phase is followed by a more complex stage of nonlinear growth that can be simplified with a model of spherical collapse and virialization, also presented here. Once the fluctuation is in virial equilibrium (forming a so-called dark matter halo), we consider the issue of gas cooling, responsible for the typical sizes and masses of galaxies. The distribution of the dark matter halos leads us to a statistical treatment based on the Gaussian distribution, following the Press-Schechter argument. This treatment results in a hierarchical buildup of structure. Finally, galaxy clustering is presented as a way of exploring the underlying cosmology.

7.1 A cosmology primer

Galaxy formation, in its earliest stages, involves the growth of small density fluctuations in an otherwise homogeneous and isotropic Universe. This
approximation, called the cosmological principle, represents the central
tenet of our current paradigm of cosmology, allowing us to use a highly
simplified metric as the solution to Einstein's equations of general relativity. This textbook avoids a relativistic treatment, as stellar and galactic
velocities are, at most, $\sim 10^3$ km s^{-1} $\ll c$, and the ratio of gravitational
to rest-mass energy is $W/Mc^2 \sim 10^{-3} M_{10}/R_{\rm kpc}$, where M_{10} is the mass in
units of $10^{10} M_\odot$, and $R_{\rm kpc}$ is the typical size of the stellar system in kpc.
However, it is necessary to invoke the spacetime metric to quantify the
(important) role of the expansion of the Universe in the growth of density
fluctuations. The Friedmann-Lemaître-Robertson-Walker metric (FLRW)
depends on only two parameters, an overall dimensionless scale factor,
$a(t)$, which accounts for the physical separation between two galaxies
with cosmic time, and the curvature constant, κ, with three different
values: $-1, 0, +1$ for a closed, flat and open Universe, respectively:

$$ds^2 = c^2 dt^2 - a^2(t) \left[\frac{dr^2}{1 - \kappa r^2} + r^2 \left(d\theta^2 + \sin^2 \theta d\phi^2 \right) \right]. \qquad (7.1)$$

The scale factor is set to unity at present time: $a(t_0) = 1$. Observations of
the angular distribution of temperature in the Cosmic Microwave Background (CMB) impose stringent constraints on the curvature, leading to
a flat Universe: $\kappa = 0$. Although we cannot measure the scale factor directly, it is possible to relate it to a direct observable: the wavelength of
a photon emitted from a distant source (say at cosmic time t). Since all
scales are affected by $a(t)$, we can write

$$\frac{\lambda(t)}{a(t)} = \frac{\lambda(t_0)}{a(t_0)}, \qquad (7.2)$$

where $\lambda(t) \equiv \lambda_0$ is the rest-frame wavelength, and $\lambda(t_0) \equiv \lambda_{\rm obs}$ is the observed wavelength. Noting that the redshift (z) is defined as: $1 + z \equiv$
$\lambda_{\rm obs}/\lambda_0$, we find that

$$a(t) = \frac{1}{1 + z}. \qquad (7.3)$$

Therefore, a galaxy at redshift $z = 1$ lives in a Universe whose size is
half of the size of the present one. We will see that in cosmology one
can interchangeably use time, the scale factor or redshift to track cosmic
evolution.

Another important parameter is the rate of change of the scale factor. It can be measured as the recession velocity between distant galaxies. Note that in order for these estimates to be valid within the context of the cosmological principle, we need to work with large enough distances so that homogeneity and isotropy hold. Therefore, the relative velocity between two nearby galaxies will be affected by local interactions (termed 'peculiar velocities'). Incidentally, regardless of the expansion of the Universe, our nearest neighbour, the Andromeda galaxy, is moving towards us! It is for much more distant galaxies that the effect of peculiar velocities is negligible, and the cosmological expansion (Hubble flow) dominates. A fundamental relation is the Hubble law, where the recession velocity (v_r) is

$$v_r = \frac{\dot{a}(t)}{a(t)} d = H(t)d, \tag{7.4}$$

and d is the separation between the galaxies. Since all galaxies are affected by the cosmic expansion, it is sometimes useful to define a comoving distance, $r_{\mathrm{com}} = d/a(t)$, thereby effectively factoring out the expansion.[1] In comoving coordinates, the only motion between two galaxies is caused by peculiar velocities, a useful framework when tracing the growth of density fluctuations. The Planck 2015 value (see table 7.1) of Hubble's constant is $H(t_0) = (67.8 \pm 0.9)\,\mathrm{km\,s^{-1}\,Mpc^{-1}}$. Noting that peculiar velocities among galaxies can be as high as $10^3\,\mathrm{km\,s^{-1}}$, we can infer that distances between galaxies much larger than $\sim 20\,\mathrm{Mpc}$ are needed to be able to 'feel' the cosmological expansion.

Friedmann's equations

Friedmann's equations provide a solution to Einstein's General Relativistic (GR) equations in the FLRW metric. Although GR is beyond the scope of this textbook, we need to consider GR to effectively link the geometry of spacetime (the metric) to the distribution of matter and energy in the system. Friedmann's equations allow us to relate the evolution of the scale factor, $a(t)$, to the (homogeneous) density in its various guises: radiation, matter or dark energy. We will show here a simplified derivation based on Newtonian mechanics which arrives at the correct equations. Let us model the Universe as a sphere with radius r, expanding with the Hubble flow. The force on a test particle with mass m is

$$m\ddot{r} = -\frac{GM(<r)m}{r^2} = -\frac{4\pi Gm}{3} r \rho(t) \Longrightarrow \ddot{a} = -\frac{4\pi G}{3} \rho(t)a(t), \tag{7.5}$$

where we have identified the radius of this 'sample Universe' as the scale factor. Since the Universe is an isolated system, conservation of mass gives $\rho a^3 = \rho_0 a_0^3 = $ constant. The subindex zero refers to the present time. Therefore, multiplying equation 7.5 by $\dot{a}$ gives

$$\dot{a}\ddot{a} = \frac{1}{2}\frac{d}{dt}\dot{a}^2 = -\frac{4\pi G}{3a^2}\left(\rho_0 a_0^3\right)\dot{a}. \tag{7.6}$$

Integrating this equation leads to

$$\dot{a}^2 = \frac{8\pi G}{3}\rho a^2 - \mathcal{K}c^2, \tag{7.7}$$

with the last term being an integration constant. In GR this term has the meaning of the constant curvature allowed by the FLRW metric. To derive the correct equation from this Newtonian approximation, we need to include the contribution to the energy from pressure. Following a simple thermodynamic argument, consider the amount of heat flowing into the Universe, with volume $\mathcal{V}$. Noting that the internal energy is $\mathcal{E} = \mathcal{V}\rho c^2$, we have

$$\Delta Q = \Delta \mathcal{E} + p\Delta\mathcal{V} = \mathcal{V}\Delta(\rho c^2) + (\rho c^2 + p)\Delta\mathcal{V}. \tag{7.8}$$

Since the Universe is a closed system, $\Delta Q = 0$, and since the volume is $\mathcal{V} \propto a^3$, we can write $\Delta\mathcal{V}/\mathcal{V} = 3\Delta a/a$. Therefore:

$$\frac{d\rho}{dt} + 3\frac{\dot{a}}{a}\left(\rho + \frac{p}{c^2}\right) = 0. \tag{7.9}$$

Taking the time derivative of equation 7.7, and using 7.9 to describe $d\rho/dt$, we arrive at the second of Friedmann's equations, namely:

$$\ddot{a} = -\frac{4\pi G}{3}a\left[\rho + 3\frac{p}{c^2}\right]. \tag{7.10}$$

However, this Newtonian example cannot describe the so-called cosmological constant (Λ). Friedmann's equations including this constant are

$$\left. \begin{aligned} \dot{a}^2 &= \frac{8\pi G}{3}\rho a^2 - \kappa c^2 + \frac{1}{3}\Lambda a^2 \\ \ddot{a} &= -\frac{4\pi G}{3}a\left[\rho + \frac{3p}{c^2}\right] + \frac{1}{3}\Lambda a \end{aligned} \right\}. \tag{7.11}$$

Dimensionless densities

The first of Friedmann's equations can be written in terms of the Hubble parameter:

$$H^2(t) = \frac{8\pi G}{3}\rho - \frac{\kappa c^2}{a^2} + \frac{\Lambda}{3}. \tag{7.12}$$

This equation introduces a density scale, the critical density, that is time dependent:

$$\rho_c(t) \equiv 3H^2(t)/8\pi G. \tag{7.13}$$

At present time, the critical density is $\rho_c(t_0) = 1.879 \times 10^{-26}h^2\,\mathrm{kg\,m^{-3}}$, or, in more convenient units, $2.775 \times 10^{11}h^2\mathrm{M_\odot}\,\mathrm{Mpc^{-3}}$, where Hubble's constant is usually defined as $H(t_0) = H_0 = 100h\,\mathrm{km\,s^{-1}\,Mpc^{-1}}$ (so, roughly, $h = 0.7$). The density is usually split between radiation (ρ_γ) and matter (ρ_m). These two components vary with a as $\rho_\gamma \propto a^{-4}$ and $\rho_m \propto a^{-3}$, respectively. The -3 exponent is a volume dilution factor as the Universe expands (from conservation of energy) whereas the -4 in the radiation field takes into account the additional loss of energy because of the redshift (note that the energy of a photon is $\epsilon_\gamma = h\nu = hc/\lambda$). We also define the equivalent, dimensionless density parameters, taking the (time-dependent) critical density as reference:

$$\left.\begin{aligned} \Omega_\gamma(t) &= \frac{\rho_\gamma(t)}{\rho_{\mathrm{crit}}(t)} \\ \Omega_m(t) &= \frac{\rho_m(t)}{\rho_{\mathrm{crit}}(t)} \end{aligned}\right\}. \tag{7.14}$$

The cosmological constant can also be associated to an additional energy component (dark energy), with density parameter:

$$\Omega_\Lambda \equiv \frac{\Lambda}{3H_0^2}. \tag{7.15}$$

Plugging these values into the Friedmann equations at present time ($a(t_0) = 1$) allows us to write the curvature as

$$\kappa = \frac{\Omega_\gamma + \Omega_m + \Omega_\Lambda - 1}{(c^2/H_0^2)}, \tag{7.16}$$

which means the Universe is flat (zero curvature) if the sum of the density of all components is the critical value. A simple case often used is the Einstein–de Sitter model, where $\Omega_\gamma = \Omega_\Lambda = 0$ and $\Omega_m = 1$, which can be

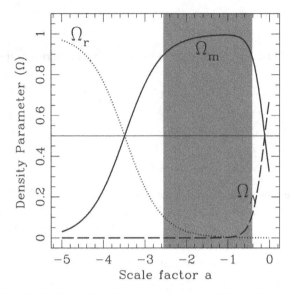

Figure 7.1 Evolution of the density parameter of the radiation (Ω_r), matter (Ω_m) and cosmological constant (Ω_Λ), as a function of the scale factor. The grey shaded area marks the region $\Omega_m \geq 0.9$, where the Universe is well described by an Einstein–de Sitter model. (Cosmological parameters taken from table 7.1.)

shown to have a simple solution for the scale factor of

$$a(t) = \left(\frac{t}{t_0}\right)^{2/3}. \tag{7.17}$$

The evolution of the density parameter is shown in figure 7.1, which illustrates the regime where an Einstein–de Sitter model gives a valid representation (grey shaded region). Note this interval corresponds to the major phase of galaxy formation.

Exercise 7.1

Show that the dimensionless matter density scales with redshift as

$$\Omega_m(z) = \frac{\Omega_{m,0}(1+z)^3}{\Omega_{m,0}(1+z)^3 + \Omega_{\gamma,0}(1+z)^4 + \Omega_{\kappa,0}(1+z)^2 + \Omega_{\Lambda,0}},$$

where the '0' subindices refer to the values at present time, and Ω_κ is the corresponding term for the curvature. In our standard

---

FUNDAMENTALS OF GALAXY DYNAMICS, FORMATION AND EVOLUTION

cosmological model (ΛCDM), the cosmological parameters are, approximately $\Omega_{m,0} \approx 0.3$; $\Omega_{\gamma,0} \approx 10^{-4}$; $\Omega_{\Lambda,0} \approx 0.7$, so that the curvature is zero. Show that during an earlier epoch, the Universe was closely represented by an Einstein–de Sitter model.

The age of the Universe

One can estimate the age of the Universe using the following integral:

$$t_U = \int_0^{t_U} dt = \int_0^{a_0=1} \frac{dt}{da} da = \int_0^1 \frac{da}{aH(a)} = \int_0^\infty \frac{dz}{(1+z)H(z)}. \qquad (7.18)$$

Note we can use the time coordinate t, the scale factor a or the redshift z to track 'cosmic time'. Theorists usually prefer the scale factor, whereas observational cosmologists prefer redshift. Applying Friedmann's equation to an Einstein–de Sitter model gives

$$H(z) = H_0(1+z)^{3/2}, \qquad (7.19)$$

leading to

$$t_U = \frac{1}{H_0} \int_0^\infty \frac{dz}{(1+z)^{5/2}} = \frac{2}{3H_0} = 6.52 h^{-1} \, \text{Gyr} \sim 9.3 \, \text{Gyr}. \qquad (7.20)$$

The most general case can be written as the equivalent of equation 7.18, using 7.12 and 7.13 to write Hubble's parameter:

$$H(z) \equiv H_0 E^{1/2}(z) = H_0\sqrt{\Omega_{m,0}(1+z)^3 + \Omega_{\gamma,0}(1+z)^4 + \Omega_{\kappa,0}(1+z)^2 + \Omega_{\Lambda,0}}, \qquad (7.21)$$

and so:

$$t_U = \frac{1}{H_0} \int_0^\infty \frac{dz}{(1+z)E^{1/2}(z)}, \qquad (7.22)$$

which, for a vanilla-flavoured ΛCDM cosmology, with $\Omega_{m,0} = 0.3$, $\Omega_{\Lambda,0} = 0.7$, $\Omega_{\gamma,0} \approx 0$, and, therefore, zero curvature, gives an age of the Universe at present of: $t_U = 9.43 \, h^{-1} \, \text{Gyr} \sim 13.7 \, \text{Gyr}$, i.e., significantly older than the Einstein–de Sitter case. To determine the age at an arbitrary redshift, z, we simply have to replace the lower limit of the integral in equation 7.18 by z.

Distance(s) in cosmology

Distance is a fundamental parameter in the interpretation of the observational properties of galaxies. In this succinct primer, we need to refer only to a fundamental difference between distance indicators. The distance directly derived from the metric (equation 7.1) is the comoving radial distance. If we consider the radial trajectory of a photon from a distant galaxy, at coordinate r, and our observing point, at $r = 0$, the distance is

$$D_c \equiv \int_0^r \frac{dr}{\sqrt{1 - \kappa r^2}} = \int_a^{a_0=1} \frac{c\,da}{a^2 H(a)} = \int_0^z \frac{c}{H(z)}\,dz, \qquad (7.23)$$

and the expression for Hubble's parameter with respect to redshift is given by equation 7.21. Note that for cosmologically small distances, we can approximate this equation by $cz \sim H_0 D_c$, i.e., Hubble's law of expansion.

However, this distance is not what we would use when measuring the angular extent of a galaxy. In this case, we measure an angle (θ) given by the ratio of its (comoving) size and the comoving distance to us. If the physical diameter of the galaxy is ø, the comoving size is $ø/a = ø(1+z)$; therefore

$$\tan\theta \approx \theta = \frac{ø(1+z)}{D_c(z)} \equiv \frac{ø}{D_a(z)}, \quad \text{where} \quad D_a(z) = \frac{D_c(z)}{1+z}, \qquad (7.24)$$

and we define the angular diameter distance D_a as the one that relates the physical size with the angular extent of the galaxy. Moreover, we also observe the flux from galaxies as the luminosity (L) per unit area. A sphere centred at the position of the observed galaxy, touching our observing position, has area $4\pi D_c^2$. However, the photon rate will decrease by a factor $a^{-1} = (1+z)$ because of the cosmological time dilation, and the energy of the photon will decrease by an additional factor of $(1+z)$ because of the cosmological redshift. Therefore, the observed flux can be written

$$f = \frac{L}{4\pi D_c^2(z)} = \frac{L_0}{4\pi D_c^2(z)(1+z)^2}$$

$$= \frac{L_0}{4\pi D_l^2(z)}, \quad \text{where} \quad D_l(z) = D_c(z)(1+z), \qquad (7.25)$$

where L_0 is the intrinsic (i.e., rest-frame) luminosity of the galaxy, and D_l is defined as the luminosity distance. Note that all three distances converge at low redshift, where large-scale effects become negligible.

7.2 Linear regime

A perfectly homogeneous Universe can be easily described by the equations presented in the previous section. However, such a density distribution would not develop structure. Hence, we need to introduce minute, but nonnegligible perturbations in the density field at early times. An inflationary scenario posits that microscopic fluctuations (over Planck scales, i.e., driven by the quantum nature of gravity) will expand at early times in an exponential way, moving outside the horizon – i.e., the region within which perturbations can be in causal contact. Being outside the horizon means the different areas of the perturbation will not be in causal contact. Therefore, these fluctuations are frozen (they cannot interact), and require general relativity (outside the scope of this book) for a correct description. Later on, because of the steady expansion, the horizon catches up with the extent of these fluctuations. For the sake of a simple, pedagogical argument, it is assumed that a standard Newtonian treatment in an expanding background suffices.

Let us consider the equations that describe the evolution of a density fluctuation as a fluid. First of all, we write down the unperturbed solution, $(v_0, \rho_0, p_0, \Phi_0)$, where we adopt the Lagrangian notation (i.e., the time derivatives follow the fluid) $d/dt = \partial/\partial t + (\vec{v} \cdot \vec{\nabla})$:

$$\left. \begin{aligned} \frac{d\rho_0}{dt} &= -\rho_0 \nabla \cdot \vec{v}_0 \\[1em] \frac{d\vec{v}_0}{dt} &= -\frac{1}{\rho_0}\vec{\nabla}p_0 - \vec{\nabla}\Phi_0 \\[1em] \nabla^2 \Phi_0 &= 4\pi G \rho_0 \end{aligned} \right\} \cdot \qquad (7.26)$$

These three expressions describe the conservation of mass, the force (i.e., Euler) equation and Poisson's equation, respectively. These equations are now modified with small perturbations in the velocity, density, pressure and gravitational potential fields: $(\delta v, \delta \rho, \delta p, \delta \Phi)$. Retaining the terms to the lowest order, we get

$$
\left.
\begin{aligned}
\frac{d}{dt}\left(\frac{\delta \rho}{\rho_0}\right) &= -\vec{\nabla} \cdot \delta \vec{v} \\[6pt]
\frac{d(\delta \vec{v})}{dt} + (\delta \vec{v} \cdot \vec{\nabla})\vec{v}_0 &= -\frac{1}{\rho_0}\vec{\nabla}(\delta p) - \vec{\nabla}(\delta \Phi) \\[6pt]
\nabla^2(\delta \Phi) &= 4\pi G \delta \rho
\end{aligned}
\right\} .
\tag{7.27}
$$

The next step involves a change of coordinates from physical distance $(\vec{x})$ to comoving distance $(\vec{r})$, factoring out the effect of the cosmological expansion:

$$
\left.
\begin{aligned}
\vec{x} &= a(t)\vec{r} \\[6pt]
\vec{v} &= \frac{d\vec{x}}{dt} = H\vec{x} + a\frac{d\vec{r}}{dt} = H\vec{x} + a\vec{u} \\[6pt]
\rho(\vec{x}, t) &= \rho_0(t)\left[1 + \delta(\vec{r}, t)\right]
\end{aligned}
\right\} .
\tag{7.28}
$$

Physical velocities $(\vec{v})$ are decomposed into a Hubble flow $(H\vec{x})$ and a peculiar velocity $(\vec{u})$. The density is also split between a smooth, background average (ρ_0) and a density contrast (δ). Note that the comoving separation between two galaxies that move only with the Hubble flow (i.e., 'at rest' in an expanding Universe) remains constant with cosmic time. Finally, we arrive at the linear perturbation equation for the density contrast:

$$
\ddot{\delta} + 2\left(\frac{\dot{a}}{a}\right)\dot{\delta} - 4\pi G \rho_0 \delta = \frac{c_s^2}{a^2}\nabla_r^2 \delta + O(\delta^2, \dot{\delta}^2).
\tag{7.29}
$$

This equation has an oscillator term $(\propto \delta)$ driven by the background matter density, a damping term caused by the expansion $(\propto \dot{\delta})$, and a driving force, on the RHS, with a characteristic scale given by the sound speed $(c_s = \sqrt{\delta p/\delta \rho})$. This equation is similar to the growth of a fluctuation in standard Newtonian dynamics (say, the collapse of a large cloud of gas in a galaxy), with the difference being the damping term caused by the expansion. This term fundamentally affects the rate of structure growth. For example, in an Einstein–de Sitter Universe $(\Omega_m = 1)$, and assuming dust-like matter $(c_s = 0)$,

$$
\begin{aligned}
a(t) &= \left(\frac{3}{2}H_0 t\right)^{2/3} \Rightarrow \frac{\dot{a}}{a} = \frac{2}{3t}, \\[8pt]
\Omega &= \frac{8\pi G \rho_0}{3H^2} = 1 \Rightarrow 4\pi G \rho_0 = \frac{2}{3t^2},
\end{aligned}
\tag{7.30}
$$

giving the following linear perturbation equation:

$$\ddot{\delta} + \frac{4}{3t}\dot{\delta} - \frac{2}{3t^2}\delta = 0. \tag{7.31}$$

If we try a solution of the type $\delta(t) \propto t^n$, we find solutions for $n = -1$ (decaying mode) and $n = 2/3$ (growing mode). Neglecting the decaying mode, we find that

$$\delta(t) \propto t^{2/3} \propto a(t). \tag{7.32}$$

Therefore, a density fluctuation will grow linearly with respect to the scale factor, in contrast with the density perturbation in a gas cloud within a galaxy, which follows an exponential rate: the cosmological expansion dampens this growth.

Exercise 7.3

Neglect the pressure term in the linear growth of density fluctuations (i.e., the term with $\nabla^2\delta$), and consider an empty Universe (i.e., zero matter and energy density). This is the Milne model. Show that in this case, density fluctuations will remain constant with time.

Jeans mass

The treatment of the density growth presented above traces only the time evolution, but it does not take into account the scale of the fluctuation. The linear density growth equation (7.29) can be applied to a specific lengthscale. One can describe the spatial distribution of the density fluctuation as a Fourier series in space,

$$\delta(\vec{r}; t) = \sum_k \delta_k(t)e^{ikr}, \tag{7.33}$$

where the coefficients, δ_k quantify the contribution of lengthscale $\lambda = 2\pi/k$ to the fluctuation. For simplicity, we will assume spherical symmetry, considering k a scalar quantity. In the linear regime, each component, δ_k, evolves independently, giving

$$\ddot{\delta}_k + 2\left(\frac{\dot{a}}{a}\right)\dot{\delta}_k = \left(4\pi G\rho_0 - c_s^2 k^2\right)\delta_k, \tag{7.34}$$

where we have used the comoving wave-number $k_c = k/a$. We can solve this equation with the ansatz

$$\delta_{k_c}(t) = \delta_{k_c,0}e^{-i\omega t},\qquad (7.35)$$

which leads to a dispersion relation

$$\omega^2 = c_s^2 k_c^2 - 4\pi G\rho_0.\qquad (7.36)$$

Hence, the small density fluctuations will oscillate if the RHS is positive. If it is negative, the fluctuations will have exponential growing/decaying behaviour. This means density fluctuations will only grow if

$$4\pi G\rho_0 > c_s^2 k_c^2 \Rightarrow \lambda > \lambda_J \equiv \frac{2\pi}{k_J} = c_s\left(\frac{\pi}{G\rho_0}\right)^{1/2}.\qquad (7.37)$$

The second part refers to the size of a (comoving) fluctuation (λ), defining the Jeans length. Any density perturbation whose size is greater than λ_J will grow. This threshold can also be given as a mass (called the 'Jeans mass'):

$$M_J = \frac{4\pi}{3}\rho_0\lambda_J^3.\qquad (7.38)$$

The Jeans instability reflects a balance between pressure forces (related to the sound speed) and gravitational forces (from ρ_0). If a fluctuation involves a mass greater than M_J, then gravity will overcome pressure, and the fluctuation will collapse. For smaller masses, sound waves will counteract the gravitational pull, preventing the formation of a structure.

Let us consider the growth of galaxies after photon decoupling. At this epoch ($z \lesssim 1{,}100$), we can make the simplifying assumption that the sound speed of the gas is given by that of monoatomic hydrogen gas:

$$c_s = \sqrt{\frac{5k_BT}{3m_P}} \sim 3.7T_3^{1/2}\,\mathrm{km\,s^{-1}},\qquad (7.39)$$

with $T_3 \equiv T/1000\,\mathrm{K}$. Close to decoupling, $T_3 \approx 3$, and the comoving Jeans length is 0.014 Mpc, leading to a Jeans mass around $1.6 \times 10^6 M_\odot$ (where we made the assumption of an Einstein–de Sitter Universe and $H_0 = 70\,\mathrm{km\,s\,Mpc^{-1}}$). Therefore, anything more massive than a globular cluster can, in principle, form a collapsed structure. However, the coupling of the ordinary matter to the photons at earlier times washed out any potential fluctuations over these scales. We will see in the next section how an additional matter component is needed to produce the high levels of density contrast we see in the Universe today.

The need for (nonbaryonic) dark matter

The linear behaviour of a perturbation implies that a long time is needed to evolve from the the early linear regime ($\delta \gg 1$) to the present distribution of matter in the Universe. Furthermore, observations of the Cosmic Microwave Background – which give a glimpse of the density fluctuations in the baryonic content of the Universe at photon-electron decoupling, i.e., 380,000 yr after the Big Bang – reveal temperature inhomogeneities around $\delta T/T \sim 10^{-5}$. Considering adiabatic temperature fluctuations, i.e., keeping the entropy, $S \propto T^3 n_\gamma^{-1}$ constant, we can write

$$\frac{\delta T}{T} = \frac{1}{3}\frac{\delta n_\gamma}{n_\gamma} \sim \frac{1}{3}\frac{\delta \rho}{\rho}. \tag{7.40}$$

It is assumed that before recombination, photons and electrons/baryons were well coupled (via Thomson scattering). Hence, the measurements of the CMB temperature fluctuations imply a density contrast at decoupling (at redshift $z_{\text{CMB}} \sim 1,000$) around $\delta(t_{\text{CMB}}) \sim 3 \times 10^{-5}$. Using a simple argument for the growth in an Einstein–de Sitter model (equation 7.32), we find those fluctuations should be at present time as follows:

$$\delta(t_0) \sim \delta(t_{\text{CMB}}) \left(\frac{a(t_0)}{a(t_{\text{CMB}})} \right) = \delta(t_{\text{CMB}})(1 + z_{\text{CMB}}) \sim 0.1. \tag{7.41}$$

This means the contrast of the baryon density field at the time of decoupling is not large enough to create galaxies today! Hence, we need an additional component, not coupled to photons, so that the growth of their density seeds took place at earlier stages, unimpeded. After decoupling, the baryons fall on to the (larger) density fluctuations of this (dark) matter field, allowing for larger values of the density contrast.

Therefore, let us consider a two-fluid component involving baryons and dark matter (the contribution from radiation is negligible at decoupling and afterwards). The growth of density fluctuations of both fluids is described by the linear perturbation equation (neglecting the pressure term):

$$\left.\begin{aligned}
\ddot{\delta}_B + 2\frac{\dot{a}}{a}\dot{\delta}_B &= 4\pi G(\rho_{0,B}\delta_B + \rho_{0,DM}\delta_{DM}) \\
\ddot{\delta}_{DM} + 2\frac{\dot{a}}{a}\dot{\delta}_{DM} &= 4\pi G(\rho_{0,DM}\delta_{DM} + \rho_{0,B}\delta_B)
\end{aligned}\right\}. \tag{7.42}$$

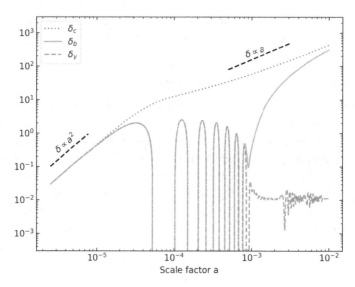

Figure 7.2 Evolution of the density contrast with cosmic time, parameterized with the scale factor, a. (Plot produced with the PYCAMB code of Lewis & Challinor, Astrophysics Source Code Library, ascl:1102.026.)

Assuming a dominant contribution from the dark matter, we find (for the Einstein–de Sitter case) that

$$\delta_{DM} = \delta_{DM}(0)a(t), \qquad (7.43)$$

which goes back into the first equation, giving

$$\ddot{\delta}_B + 2\left(\frac{\dot{a}}{a}\right)\dot{\delta}_B = 4\pi G a \rho_{0,DM}\delta_{DM}(0). \qquad (7.44)$$

Rewriting the differential equation with respect to the scale factor (or redshift), instead of time, we find that

$$\delta_B(z) = \delta_{DM}(z)\left(1 - \frac{1+z}{1+z_N}\right), \qquad (7.45)$$

where z_N is some reference value (obtained from the constant of integration). Hence, at high redshift, during the earliest phases of evolution, $z \to z_N$, and the baryon fluctuations are negligible (e.g., during the epoch of recombination/decoupling, $z_N \sim 1,000$), whereas later on, at $z \ll 1,000$, the density fluctuations of the baryons follow the dark matter distribution, $\delta_B \sim \delta_{DM}$. Figure 7.2 shows the evolution of the density contrast of the dark matter (δ_c), baryons (δ_b) and radiation (δ_γ) as a function of the scale factor. A standard ΛCDM model is assumed, with a scale $k = 0.3\,\mathrm{Mpc}^{-1}$. Note that all fields evolve similarly during the radiation-dominated phase (a $\delta \propto a^2$ trend is shown as reference). At later times, the photon/baryon mix undergoes a set of oscillations, and at decoupling ($a \sim 10^{-3}$), the baryon matter follows the evolution of the cold dark matter density fluctuations. At later times, the trend $\delta \propto a$ is also shown as reference.

7.3 Spherical collapse

The spherical collapse model is the simplest way to follow the evolution of a density fluctuation from its initial growth in the linear regime ($\delta \ll 1$) – where it 'detaches' from the Hubble flow – to the final state of a virialized, stable structure (i.e., a dark matter halo). We begin with a spherical region, with radius $R(t)$, evolving within an expanding background. The force equation drives the evolution of the region:

$$\frac{d^2R}{dt^2} = -\frac{GM}{R^2} = -\frac{4\pi G}{3}\rho_0(1+\delta)R. \tag{7.46}$$

Notice the analogy of this equation with the evolution of the scale factor in an expanding Universe:

$$\frac{d^2a}{dt^2} = -\frac{GM}{a^2} = -\frac{4\pi G}{3}\rho a. \tag{7.47}$$

Originally, the fluctuation is so small that the region expands with the background, but the slight overdensity gradually slows it down. At some time, t_{MAX}, the sphere will halt its expansion and start collapsing (turnaround). Conservation of energy gives

$$\frac{1}{2}\left(\frac{dR}{dt}\right)^2 - \frac{GM}{R} = \mathcal{E}(M) = -\frac{GM}{R_{\mathrm{MAX}}}. \tag{7.48}$$

The time evolution is therefore

$$t = \int \left[2GM\left(\frac{1}{R} - \frac{1}{R_{\mathrm{MAX}}}\right)\right]^{-1/2} dR. \tag{7.49}$$

This equation can be readily solved if we define $R = R_{\mathrm{MAX}}\cos^2\alpha$. Finally, if we redefine the angular parameter $\alpha = \pi/2 - \theta/2$, so that $R = 0$ at $\theta = 0$ and

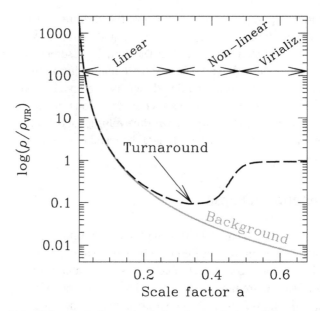

Figure 7.3 Schematic figure depicting the evolution of the density in a fluctuation following the spherical collapse model.

$R = R_{\mathrm{MAX}}$ at $\theta = \pi/2$, we arrive at the following parametric equations:

$$
\left.\begin{aligned}
R(\theta) &= R_0(1 - \cos\theta) \\
t(\theta) &= t_0(\theta - \sin\theta) \\
\rho(\theta) &= \rho_0(\theta)\frac{9(\theta - \sin\theta)^2}{2(1 - \cos\theta)^3}
\end{aligned}\right\},
\tag{7.50}
$$

which describe the trajectory of a cycloid. Note that the density equation takes into account the evolution of the background density in an Einstein–de Sitter model, i.e., $\rho_0 \propto a^{-3}$ and $a \propto t^{2/3}$. In these equations, $t_0 = \sqrt{R_{\mathrm{MAX}}^3/8GM} \propto \sqrt{1/G\rho}$, i.e., the dynamical timescale (see section 1.3). Therefore, denser fluctuations – on average found at earlier times – will have shorter collapse timescales.

In this model (see figure 7.3), the fluctuation starts expanding with the background. However it gradually slows down until it stops expanding at the turnaround time, when $\theta = \pi$. At this time, the radius of the fluctuation is $R_{\mathrm{MAX}} = 2R_0$. The density at turnaround is $5.55\rho_0$. The spherical collapse model allows us to determine the epoch at which a structure collapses (virializes):

$$\frac{1+z_{MAX}}{1+z_{VIR}} = \frac{a(2t_{MAX})}{a(t_{MAX})} = 2^{2/3} \simeq 1.59, \tag{7.51}$$

(for an EdS model). Hence, if turnaround occurs at $z_{MAX} = 10$, the system is virialized at $z_{VIR} \sim 6$. We can also probe the properties of the virialized halo by applying energy conservation:

$$E(t_{MAX}) = E(t_{VIR}) \Rightarrow -\frac{3GM^2}{5R_{MAX}} = \frac{W_{VIR}}{2} = -\frac{1}{2}\left(\frac{3GM^2}{5R_{VIR}}\right) \Rightarrow$$

$$\Rightarrow R_{VIR} = \frac{1}{2}R_{MAX}. \tag{7.52}$$

leading to a halo density at collapse of

$$\rho_{VIR} = 8\rho(t_{MAX}) \simeq 44.4\rho_0(t_{MAX}) =$$

$$= 44.4\left(\frac{1+z_{MAX}}{1+z_{VIR}}\right)^3 \rho_0(t_{VIR}) \simeq 180\rho_0(t_{VIR}), \tag{7.53}$$

providing a relationship between the density of a halo and its cosmic epoch of collapse:

$$\rho_{VIR} \sim 180\rho_{crit,0}\Omega_{m,0}(1+z_{VIR})^3. \tag{7.54}$$

For instance, for the properties of the Milky Way ($M_{DM} \sim 3 \times 10^{11} M_\odot$; $R_{VIR} \sim 50\,\text{kpc}$) we find a virialization epoch of $z_{VIR} \sim 3$, whereas the collapse of a galaxy cluster ($M_{DM} \sim 10^{15} M_\odot$; $R_{VIR} \sim 1.5\,\text{Mpc}$) happens at a later epoch $z_{VIR} \sim 1$. This result reflects the hierarchical nature of structure formation, with small, denser fluctuations occuring first. However, this result concerns the formation of dark matter halos. The galaxies are actually formed through the subsequent collapse of baryonic material (H, He gas) into the centres of these halos. Unfortunately, a direct mapping of dark matter formation and galaxy formation is not straightforward, and requires a proper understanding of processes involving, among others, the collapse of the infalling gas and its cooling, star formation and potential feedback effects. These complex mechanisms are loosely termed 'baryon physics'. Moreover the formalism assumes no mass mergers after the collapse of the halo. Nevertheless, it provides powerful insight about the relation between the epoch of formation and the properties of the dark matter halos.

Exercise 7.4

A dark matter halo forms in an Einstein–de Sitter universe, collapsing at redshift z_{vir}, with a mass density profile given by

$$\rho(r) \equiv \begin{cases} \dfrac{v_c^2}{4\pi G r^2}, & r \leq r_0, \\[2mm] 0, & r > r_0, \end{cases}$$

where r_0 represents the radial extent of the halo, and v_c is the velocity of any circular orbit in the halo. Show that the radial extent of the halo is

$$r_0 = \frac{v_c}{10 H(z_{vir})},$$

where $H(z_{vir})$ is the Hubble parameter at virialization. Assuming that the stars in a galaxy have the same circular velocity as its dark matter halo, how could we compare the formation epochs of two galaxies with the same v_c?

The linear density equivalent

In the next section we will extend spherical collapse to a statistical derivation of the halo mass function. First, however, it will be useful to derive the equivalent density contrast for collapse in the linear regime, i.e., assuming that the difference between the density of the clump and the background density can be expressed to the lowest order. In this case it is useful to rewrite equation 7.50 with respect to the turnaround radius (R_{MAX}) and time at turnaround (t_{MAX}) as

$$\left. \begin{aligned} s &\equiv \frac{R}{R_{MAX}} = \frac{1}{2}(1 - \cos\theta) \\[3mm] \tau &\equiv \frac{t}{t_{MAX}} = \frac{1}{\pi}(\theta - \sin\theta) \end{aligned} \right\} . \qquad (7.55)$$

In the linear regime we expand to the lowest nontrivial order in θ:

$$s \simeq \frac{\theta^2}{4}\left[1 - \frac{\theta^2}{12}\right], \qquad \tau \simeq \frac{\theta^3}{6\pi}\left[1 - \frac{\theta^2}{20}\right].$$

Writing out s as a function of τ by retaining the lowest order terms in θ leads to

$$s \simeq \frac{(6\pi\tau)^{2/3}}{4}\left[1 - \frac{(6\pi\tau)^{2/3}}{20}\right]. \qquad (7.56)$$

Note that the lowest order term gives the evolution of the size of the perturbation if it is moving with the Hubble flow (i.e., following an Einstein–de Sitter model). Now we write the density contrast in the linear approximation as $1 + \delta_{\text{lin}} = s_{\text{back}}/s$, where the numerator corresponds to this lowest order behaviour that follows the background expansion. Retaining again the lowest order term leads to

$$\delta_{\text{lin}} = \frac{3}{20}(6\pi\tau)^{2/3} = 1.69, \qquad (7.57)$$

where the numerical solution corresponds to the linear prediction at collapse time, i.e., $\tau = 2$.

Exercise 7.5

Numerical simulations of structure formation suggest that dark matter halos follow a universal profile (Navarro-Frenk-White, or NFW):

$$\rho\left(x \equiv \frac{r}{r_S}\right) = \frac{\delta_0 \rho_{\text{crit}}}{x(1+x)^2},$$

(assume spherical symmetry), where r_S is the scale radius of the halo, ρ_{crit} is the critical background density, and δ_0 is the density contrast. The concentration of the halo is defined as $c = r_{\text{VIR}}/r_S$. Following the spherical collapse model, show that

$$\delta_0 = \frac{180}{3}\frac{c^3}{\ln(1+c) - c/(1+c)}.$$

7.4 Press-Schechter formalism

The combination of the linear density growth equation (7.29) and spherical collapse (7.54) allows us to trace the evolution of single density perturbations from the small seeds with $\delta \ll 1$ all the way to the formation of a stable halo. The next step involves dealing with the whole distribution of density fluctuations as a random field, identifying the formation of halos in a statistical way. Such a treatment will enable us to define a mass function of virialized halos, much in the same way that we defined the luminosity function of galaxies in chapter 1. However, this process gets complicated as structures can form within structures: a galaxy in a cluster can be considered part of the dark matter halo that formed the progenitor galaxy early on, or part of the cluster that incorporated the galaxy/halo system at later times (cloud-in-cloud problem). We need a way to associate a density fluctuation that eventually forms a halo with a (mass or size) scale. To do that, the density field is convolved with a spherical top-hat filter, defining at each point the smoothed average density within a radius R_M as

$$\delta_s(\vec{r}; R_M) = \int \delta(\vec{r}')W(\vec{r} - \vec{r}'; R_M)d^3r', \qquad (7.58)$$

where the function $W(\vec{r}; R_M)$ is constant inside $|\vec{r}| \leq R_M$, and zero outside. A given scale R_M corresponds to a mass $M = 4\pi \langle \rho \rangle R_M^3/3$. The Press-Schechter formalism assumes that the probability that $\delta_s > \delta_c$ – i.e., the density contrast for collapse – is the same as the fraction of structures present at time t, contained in halos with mass greater than M. The linear theory equivalent of the density of a collapsed structure shown in equation 7.54 corresponds to $\delta_c = 1.69$ (equation 7.57). The density contrast, $\delta(\vec{r})$, can be described by a random Gaussian field, and likewise for the smoothed version, so that the probability of having a structure of mass greater than M is

$$\mathcal{P}[> \delta_c] = \frac{1}{\sigma(M)\sqrt{2\pi}} \int_{\delta_c}^{\infty} e^{-\delta_s^2/2\sigma^2(M)} d\delta_s \propto F(> M), \qquad (7.59)$$

where the variance of the Gaussian distribution is

$$\sigma^2(M) = \langle \delta_s^2(\vec{r}; R_M) \rangle, \qquad (7.60)$$

and $F(> M)$ is the fraction of halos with mass greater than M. A problem in this formalism is that as $M \to 0$, the mass variance $\sigma^2(R)$ diverges, and the total probability $\mathcal{P} \to 1/2$, as if only half of the total mass would eventually collapse into halos. Therefore, only the mass contained in overdense regions, i.e., above average, are assumed to collapse into halos. However, underdense regions can form part of collapsed structures, and Press and Schechter argued,[2] without proof, that all matter will eventually collapse into halos, therefore requiring an extra factor of 2 to account for this: $F(> M) = 2\mathcal{P}[> \delta_c]$, so that the number density of halos with masses in the range M and $M + dM$ is

$$n(M)dM = \frac{\langle \rho \rangle}{M} \frac{\partial F(> M)}{\partial M} dM = \sqrt{\frac{2}{\pi}} \frac{\langle \rho \rangle}{M^2} \frac{\delta_c}{\sigma} e^{-\delta_c^2/2\sigma^2} \left| \frac{d \ln \sigma}{d \ln M} \right| dM. \quad (7.61)$$

The Press-Schechter formalism gives a useful tool to map the evolution of structure in a statistical way. Only halos with mass M form in a significant number if $\sigma(M) \gtrsim \delta_c$. The time evolution of the mass function is implicit in the variation of $\sigma(M)$ with cosmic time, defined as $\sigma(M, t) = \sigma(M, 0)D(t)$, where $D(t)$ is the growth factor of fluctuations, for instance in the Einstein–de Sitter case, $D(t) = (t/t_0)^{2/3}$. An alternative way to introduce the time evolution is to adopt a different density contrast for collapse with cosmic time, following $\delta_c(t) = \delta_c(0)/D(t)$. In the standard ΛCDM framework, $\sigma(M)$ decreases with mass, implying that halos form in a bottom-up way, from small, dense structures at early times towards massive halos at later times. If we assume a simple scaling relation for the variance,

$$\sigma = \left(\frac{M}{M_0} \right)^{-\alpha}, \quad (7.62)$$

we obtain the following halo mass function:

$$n(M) = n_0 \left(\frac{M}{M_\star} \right)^{\alpha - 2} e^{-\left(\frac{M}{M_\star} \right)^{2\alpha}}, \quad (7.63)$$

where n_0 is the normalization factor, and

$$M_\star = M_0 \left(\frac{2}{\delta_c^2} \right)^{1/2\alpha} \quad (7.64)$$

is the characteristic mass scale. Note this result is reminiscent of the Schechter function (see equation 1.5), presented in the description of the luminosity function as a power law with an exponential cutoff. As cosmic time increases, the density contrast threshold decreases as $1/D(t)$, pushing the exponential cutoff towards higher masses, as expected in the progressive growth of structure.

7.5 Correlation function

In addition to the census of halos with respect to mass, given by the Press-Schechter formalism, one can also consider the correlation of the density fluctuations: Are they randomly distributed in space, or do they preferentially cluster? The clustering information provides a strong discriminant among different structure formation scenarios. We quantify this property with the two point auto-correlation function $\xi(\vec{r})$, defined as the excess probability of finding a pair of galaxies (or clumps, or halos) separated by a radius vector $\vec{r}$, with respect to a purely homogeneous distribution. The number of galaxy pairs within infinitesimal volumes dV_1 and dV_2 is therefore

$$dN_p(\vec{r}) = N_0^2[1 + \xi(\vec{r})]dV_1 dV_2, \tag{7.65}$$

where N_0^2 gives the value for a homogeneous, uncorrelated distribution. This can be related to the density contrast:

$$dN_p(\vec{r}) = \rho(\vec{x})dV_1\rho(\vec{x}+\vec{r})dV_2 = \rho_0^2[1+\delta(\vec{x})][1+\delta(\vec{x}+\vec{r})]dV_1 dV_2. \tag{7.66}$$

Averaging over a large number of volume elements,

$$dN_p(\vec{r}) = \rho_0^2[1 + \langle\delta(\vec{x})\delta(\vec{x}+\vec{r})\rangle]dV_1 dV_2. \tag{7.67}$$

Hence, the two-point correlation function is

$$\xi(r) = \langle\delta(\vec{x})\delta(\vec{x}+\vec{r})\rangle. \tag{7.68}$$

Invoking isotropy, this function is assumed to depend only on distance but not on direction. The observed correlation function is found to be well represented by a simple power law, such that

$$\xi(r) \propto \left(\frac{r}{r_0}\right)^{-\gamma} \tag{7.69}$$

over physical scales between $100\,h^{-1}\text{kpc}$ and $10\,h^{-1}\text{Mpc}$, with a characteristic length scale $r_0 = 5\,h^{-1}\text{Mpc}$, and $\gamma = 1.8$. This trend depends on galaxy morphology, with elliptical galaxies being more clustered than disc galaxies. The connection between the correlation function and the power spectrum can be established if we take the Fourier transform of the density contrast:

$$\delta(\vec{r}) = \frac{V}{(2\pi)^3}\int \delta_{\vec{k}}e^{-i\vec{k}\cdot\vec{r}}d^3k,$$
$$\delta_{\vec{k}} = \frac{1}{V}\int \delta(\vec{r})e^{i\vec{k}\cdot\vec{r}}d^3r. \tag{7.70}$$

Using Parseval's theorem relating Fourier transform pairs, we find that

$$\frac{1}{V} \int \delta^2(\vec{r}) d^3 r = \frac{V}{(2\pi)^3} \int |\delta_{\vec{k}}|^2 d^3 k, \qquad (7.71)$$

where the mean square of the density fluctuations (LHS= $\langle \delta^2 \rangle$) is given with respect to the quantity $|\delta_{\vec{k}}|^2$, defined as the power spectrum $P(k)$. For each value of k, the power spectrum gives the 'intensity' of density fluctuations over lengthscales $\lambda = 2\pi/k$. Using the definition of the autocorrelation function from equation 7.68, we get

$$\xi(r) = \frac{V}{(2\pi)^3} \int P(k) e^{i \vec{k} \cdot \vec{r}} d^3 k. \qquad (7.72)$$

Hence, the two-point autocorrelation function and the power spectrum are Fourier pairs. Assuming spherical symmetry, and inverting the previous integral, we can compute the power spectrum from observations of galaxy clustering:

$$P(k) = \frac{1}{V} \int_0^\infty \xi(r) \frac{\sin kr}{kr} 4\pi r^2 dr. \qquad (7.73)$$

The primordial power spectrum is assumed to be a simple power law:

$$P_0(k) \propto k^n; \qquad (7.74)$$

this implies a correlation function (for the primordial structures) of

$$\xi_0(r) \propto \int \frac{\sin kr}{kr} k^{n+2} dk \sim r^{-(n+3)} \propto M^{-(n+3)/3}, \qquad (7.75)$$

where we have transformed the distance r into a mass scale M via $M \sim \rho r^3$. We can relate this scaling to the density fluctuations (eq. 7.68) over a mass M:

$$\delta_0(M) \equiv \langle \delta^2 \rangle^{1/2} \propto M^{-(n+3)/6}. \qquad (7.76)$$

At very early times, fluctuations grow as $\delta \sim a^2(t)$ (see figure 7.2), and the amount of mass within the horizon goes as $M \propto a^3 \Rightarrow a \propto M^{1/3}$. Hence, the density contrast of fluctuations over a mass scale M at early times – when they cross the horizon – is

$$\delta(M, t) \propto a^2(t) \delta_0(M) \propto M^{2/3} M^{-(n+3)/6} \propto M^{-(n-1)/6}. \qquad (7.77)$$

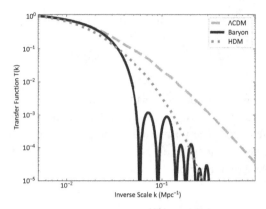

Figure 7.4 Transfer function for three different cosmological models: a standard ΛCDM scenario (see table 7.1); a baryon-dominated model where the Ω_b and Ω_m values of the ΛCDM model are swapped; and a hot dark matter (HDM) model, dominated by massless neutrinos. (Plot produced with the PYCAMB code of Lewis & Challinor, Astrophysics Source Code Library, ascl:1102.026.)

Hence, for $n = 1$, the fluctuations are independent of mass. This is a desirable behaviour that defines the Harrison-Zeldovich spectrum. The primordial density fluctuations will be affected by a number of physical processes, removing power from specific lengthscales. For instance, hot dark matter removes power from small scales, because of free streaming. The transfer function $T(k)$ relates the original fluctuations to the evolved ones, namely:

$$\delta_k(t) = \delta_k(0)T(k)D(t), \qquad (7.78)$$

where $D(t)$ is the growth of a fluctuation with cosmic epoch, derived from the geometry of spacetime, e.g., for an EdS universe $D(t) = a(t) = t^{2/3}$ (see equation 7.32). Some of the effects from the transfer function can be seen directly in the two-point autocorrelation function. The wiggles at scales $\sim 100h^{-1}$ Mpc (related to the evolution of the growth function with time, as shown in figure 7.4) reflect the Baryon Acoustic Oscillations. They are a very powerful tool to constrain cosmological models.

Types of dark matter and the power spectrum

The fundamental aspect that defines the role of dark matter in the formation and evolution of structure is the average energy per particle at decoupling, i.e., when the expansion rate exceeds the interaction rate of dark matter particles with the rest of the components, much in the same way as decoupling between the photon and electron/proton components, but at earlier times. We can differentiate three types of dark matter that will leave different imprints on the density distribution:

Hot dark matter (HDM) is made of particles that are relativistic at decoupling. They are low-mass (but non-zero), fast moving particles (e.g., standard neutrinos) that do not cluster into small structures, because a large mass ($> 10^{14} M_\odot$) is needed to keep them gravitationally bound. Since the speed of sound is close to the speed of light, the Jeans mass must be very high. Structure formation evolves from the top-down, i.e., starting with the formation over large scales/masses, with fragmentation later producing galaxy-type structures.

Cold dark matter (CDM) consists of particles that are nonrelativistic at decoupling. They clump over all scales, leading to a bottom-up process of structure formation. Examples are heavy neutrinos, axions (these ones may be nonthermal relics), or the more generic WIMPS (Weakly Interacting Massive Particles), such as the neutralino (involving supersymmetry).

Warm dark matter (WDM) has intermediate properties between HDM and CDM. The formation of small structures is suppressed, perhaps solving an excess of small halos in a CDM scenario. However, this also delays the formation process of galaxies, which is at odds with the observation of galaxies with very old stellar populations at present time, and the presence of massive galaxies at relatively high redshift.

As the growth of density fluctuations as a function of mass depends on the nature of dark matter, the measurement of the density fluctuation power spectrum at a given cosmic epoch will constrain the nature of dark matter. The scale-dependent growth of fluctuations is described by the Transfer Function, shown in figure 7.4 for three different cosmological models. Note that the HDM model will suppress density fluctuations over small scales, with respect to the standard ΛCDM framework. The baryon-dominated scenario also suppresses fluctuations over small scales, because of its early coupling with the photon fluid (indeed, justifying the need for nonbaryonic dark matter; see section 7.2), and introduces a significant amount of oscillations.

CDM is the favoured model – along with a cosmological constant, defining the ΛCDM model. The Planck 2015 best-fit cosmological parameters (see table 7.1) show that only $\approx 1/6$ of the matter content in the Universe is in the form of 'baryons'. You should also contrast this with the much lower fraction ($\sim$3%) expected in galaxies (figure 1.6), suggesting that a large amount of baryons was ejected from halos and should live in the intergalactic medium, in the form of a diffuse gas. Figure 7.1 shows

Table 7.1 Cosmological parameters.

Parameter	Symbol	Value
Matter density	Ω_m	0.3089 ± 0.0062
Baryon density	Ω_b	0.0486 ± 0.0010
Dark energy density	Ω_Λ	0.6911 ± 0.0062
Hubble constant	H_0	67.74 ± 0.46
Power spectrum normalizarion	σ_8	0.8159 ± 0.0086
Power spectrum index	n	0.9667 ± 0.0040

Source: Planck collaboration, 2016, A&A, 594, 13.

Note: All parameters are dimensionless, except for the Hubble constant, given in $\mathrm{km\,s^{-1}\,Mpc^{-1}}$.

that the expansion at present is dominated by the cosmological constant (dark energy).

7.6 Cooling and the masses of galaxies

Right after the formation (virialization) of a dark matter halo, we can assume that the gas is homogeneously distributed in this halo, at the virial temperature. However, the gas will lose energy, settling into the central region, cooling down and eventually forming stars. A very simple argument, invoking the cooling function, leads to the physical properties of structures where galaxies form. The cooling function, $\Lambda(T,Z)$, is defined such that the energy loss per unit time of gas with (electron) number density n is $dE/dt = n^2 \Lambda(T,Z)$. Hence, one can define a cooling timescale as

$$t_{\mathrm{cool}} \equiv \frac{E}{\dot{E}} = \frac{\frac{3}{2} nkT}{n^2 \Lambda(T,Z)}. \tag{7.79}$$

The cooling function is governed by processes that result in emission of photons, with the subsequent loss of energy from the system. This complex derivation is simplified by assuming a balance between the collisional ionization and recombination of the different atom and ion species. When a galaxy forms, two fundamental timescales are confronted: the cooling timescale and the dynamical timescale, where the latter depends only on the mass density. The collapse time of a homogeneous sphere with mass M and radius R is

$$t_{\mathrm{dyn}} \approx \frac{1}{\sqrt{G\rho}} \sim 5 \times 10^7 \, \mathrm{yr} \left(\frac{n}{1\,\mathrm{cm}^{-3}} \right)^{-1/2}. \tag{7.80}$$

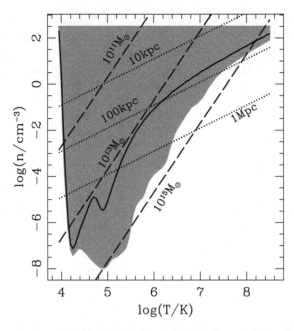

Figure 7.5 Region where galaxy formation is allowed, following the Rees-Ostrikercriterion.[a] The solid and dashed lines delimit the parameter region corresponding to fixed values of mass and size, respectively.

---

[a]Rees & Ostriker 1977, MNRAS, 179, 541.

The difference between dynamical, cooling (and cosmological) time-scales determines whether a baryon gas can form a galaxy.[3] There are three main regimes to consider:

1. $H_0^{-1} < t_{\rm cool}$: No structure can form, and the gas stays as a diffuse component.

2. $H_0^{-1} > t_{\rm cool} > t_{\rm dyn}$: A hot structure can be formed, but the cooling is not efficient enough, forming a hot, pressure-supported gaseous halo, as in the intracluster medium (see chapter 9).

3. $H_0^{-1} > t_{\rm dyn} > t_{\rm cool}$: Cooling proceeds, triggering the collapse of the gas in the central region of the halo, leading to galaxy formation.

Therefore, the parameter that controls galaxy formation is $\mathcal{R} \equiv t_{\rm cool}/t_{\rm dyn}$. Only for $\mathcal{R} < 1$ can the gas cool and form stars. This can be directly related to the typical sizes and masses of galaxies. Figure 7.5 shows the galaxy formation regime within the grey shaded region – inside which $\mathcal{R} < 1$, representing gas with solar chemical composition. The black wiggly line – corresponding to zero metallicity – would delimit instead a narrower region for star formation, as expected, since a higher metallicity leads to a more efficient cooling process. The dashed and dotted lines mark the trends of fixed mass and radius, respectively, assuming an ideal gas, and a 10 per cent contribution in baryons.

Exercise 7.6

Show that the slope of the lines in the log-log plot of figure 7.5 for structures with constant radius have slope 1, and the lines at constant mass have slope 3. Adopt an ideal gas equation of state and impose hydrostatic equilibrium to find the answer.

Notes

1 Note that the FLRW metric (equation 7.1) is defined with comoving coordinates.
2 Press and Schechter, 1974, ApJ, 187, 425.
3 Rees & Ostriker, 1977, MNRAS, 179, 541.

8
Smaller stellar systems: Stellar clusters

While this book focuses on galaxies, there exist smaller gravitating systems in which very interesting dynamical mechanisms that are not found in galaxies play an important role. Stellar clusters, which inhabit galaxies themselves, come in two flavours: open clusters and globular clusters. This chapter explores the properties of these systems and the differences with respect to the dynamics of a galaxy. Strong tidal forces are especially important, leading to the destruction of lower-mass clusters. In addition, the high densities found in the cores of globular clusters lead us to explore in more detail the connection between collisionless dynamics and thermodynamics. As the velocities of stars in a globular cluster evolve towards a Maxwell-Boltzmann distribution, a small but sizeable fraction of stars acquires high speeds and leaves the system. This process leads to a slow but gradual evaporation. The King models are a simple way of describing these systems. In the limit of very high densities, a collisional, gravitating structure would collapse into a singularity (gravothermal catastrophe). However, we do not find massive black holes at the centres of globular clusters. Other mechanisms counteract this collapse, most notably the formation of tightly bound binaries (Heggie's law).

8.1 Open and globular clusters

Stellar clusters are sites of past star formation activity, where gas clouds cooled and fragmented. Their dynamical evolution is strongly affected by their motion in the Galaxy. The different morphological appearances of open and globular clusters (figure 8.1) reflect a substantially

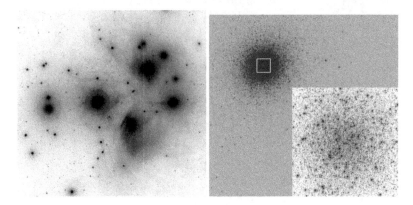

Figure 8.1 Image of an open cluster (*left*, Pleiades) and a globular cluster (*right*, M13). DSS-2 images, and inset of M13 taken by the Hubble Space Telescope/ACS. (Source: courtesy Space Telescope Science Institute.)

disparate formation scenario. Globular clusters (GCs) are very dense stellar systems, where up to a million stars can be found within a spherical volume of a few parsec in radius. Their stellar populations are overall very old and metal poor. The orbits of GCs are not confined to the plane of the Galaxy, and are instead found over a large spheroidal volume. These properties suggest an early formation process during the collapse of the gas component towards the centre of the dark matter halo. Our Milky Way galaxy contains about 200 GCs, with half of them located within 10 kpc of the Galactic centre, but with a number of them being very distant ($\gtrsim 50$ kpc).

Exercise 8.1

Consider the motion of a globular cluster of mass m moving along a circular orbit around a galaxy with an isothermal potential $\rho(r) = v_c^2/4\pi G r^2$, where v_c is the (constant) orbital velocity. Dynamical friction (see section 3.10) will slow down the orbit. Noting that this force acts tangentially along the orbit, find the rate of decrease of angular momentum and the time it would take to sink to the centre of the galaxy from some initial distance r_0. Assume constant orbital velocity throughout.

In contrast, open clusters (OCs) comprise younger and more metal rich populations, and their orbits are mostly confined to the galactic plane. Moreover, they are less dense systems than GCs, and there is an age-compactness relation, so that older OCs are more dispersed. These properties show that OCs are typical sites of star formation in the Galaxy. In the past, about 4.5 Gyr ago, the Sun formed as part of an open cluster. Young systems such as the Orion nebula provide a glimpse into the very formation of an OC.

8.2 Internal evolutionary effects

We will follow a simple argument based on the gravitational energy content of stellar clusters to explore the evolution of these systems under the effect of internal mechanisms. Starting from the virial theorem,

$$W = 2E = -\gamma \frac{GM^2}{R}, \tag{8.1}$$

and assuming γ does not evolve, i.e., the stars in the cluster keep the same mass distribution, we find that

$$\frac{dR}{R} = 2\frac{dM}{M} - \frac{dE}{E}. \tag{8.2}$$

Let us examine the effects of two important mechanisms.

Mass loss by stellar winds

Stellar winds are processes linked to stellar evolution, where the outer layers of stars are being pushed away, for instance in massive Wolf-Rayet stars, or in the later evolutionary phases (e.g., Asymptotic Giant Branch) of lower mass stars. We could also include here the mass ejection from supernovae explosions, as massive stars reach their endpoints. If a mass dM escapes from the cluster, there is a net change in kinetic energy:

$$dT = \frac{1}{2}dM \left(\langle v_e^2 \rangle - \langle v^2 \rangle \right) = \frac{3}{2} \langle v^2 \rangle dM = -dE. \tag{8.3}$$

The last steps involve the virial theorem. Using $\langle v^2 \rangle = -2E/M$,

$$\frac{dE}{E} = 3\frac{dM}{M} \Rightarrow \frac{dR}{R} = -\frac{dM}{M} \Rightarrow R \propto 1/M. \tag{8.4}$$

Consequently, mass loss leads to cluster expansion, by a factor of two or more during the lifetime of a cluster. If the mass loss takes place very quickly – for example, as a result of young hot stars blowing away the pro-tostellar material – then we have an energy $E_0 = T_0 + W_0 = M_0 \langle v^2 \rangle /2 - \gamma GM_0^2/R$, before the event, and $E = \beta T_0 + \beta^2 W_0$, after the event, where β is the mass fraction remaining in the cluster (we assume the mass loss is so rapid that both $\langle v^2 \rangle$ and R do not have time to adjust). Hence

$$E = \beta(2\beta - 1)E_0. \tag{8.5}$$

If half of the mass were lost, we total energy $E = 0$, which means the cluster becomes unbound. This is one of the mechanisms that explain why so few stars (<1 per cent in the Galactic disc) live in open clusters.

Core collapse: The gravothermal catastrophe

Consider a cluster where evolution is self-similar (i.e., the shape and mass distribution do not change). Invoking once more the virial theorem, we can write:

$$E = -\frac{\gamma}{2} \frac{GM^2}{R}, \tag{8.6}$$

where γ is a dimensionless parameter that depends on the density profile. If mass loss is caused only by evaporation, we can neglect the change in energy. Hence

$$-\frac{\gamma}{2} \frac{GM_0^2}{R_0} = -\frac{\gamma}{2} \frac{GM^2}{R} \Rightarrow R = R_0 \left(\frac{M}{M_0} \right)^2. \tag{8.7}$$

The mass loss rate can be written

$$\frac{dM}{dt} = -\lambda \frac{M}{t_{\text{rel}}}, \tag{8.8}$$

where λ is the mass fraction loss due to evaporation per unit relaxation time (i.e., $\lambda \lesssim 0.01$). Here we need to know only the scaling of the relaxation time with respect to the mass and the radius of the cluster. From equation 3.23,

$$t_{\text{rel}} \propto \langle v^2 \rangle^{3/2}/\rho \propto \sqrt{MR^3},$$

and using equation 8.7,

$$t_{\text{rel}} = t_{\text{rel},0} \left(\frac{M}{M_0} \right)^{7/2},$$

we can integrate equation 8.8, getting

$$M(t) = M_0 \left(1 - \frac{7}{2} \frac{\lambda t}{t_{\mathrm{rel},0}} \right)^{2/7}. \tag{8.9}$$

Therefore, the cluster will evaporate in a finite time given by

$$\frac{7}{2} \frac{\lambda t_{\mathrm{ev}}}{t_{\mathrm{rel},0}} \sim 1 \Rightarrow t_{\mathrm{ev}} \sim \frac{2 t_{\mathrm{rel},0}}{7 \lambda} \sim 40 t_{\mathrm{rel},0},$$

where the last estimate uses $\lambda \sim 0.7\%$. Turning now to the evolution of the stellar density in the cluster,

$$\rho \propto \frac{M}{R^3} \propto M^{-5},$$

and using equation 8.9, we find that

$$\frac{d\rho}{dt} \propto \left(1 - \frac{7}{2} \frac{\lambda t}{t_{\mathrm{rel},0}} \right)^{-17/7}.$$

The density increases very quickly with time, leading to core collapse. Tidal forces will accelerate the evaporation rate. This runaway process is inherently caused by the properties of the gravitational force. If we model the stars in a cluster as a gas of particles, in analogy with statistical mechanics, we find that this gas features a negative heat capacity (as 'temperature' increases, the internal energy gets more negative). Therefore, as the system 'loses' thermal energy, it gets hotter, i.e., the velocity dispersion rises. The system will evolve away from an isothermal distribution, towards a state of higher entropy consisting of an isothermal core and a nonisothermal envelope. The core collapses, becoming more tightly bound (a cuspy structure), while the halo expands. A singularity may be found at the centre. Roughly 20 per cent of globular clusters have central density profiles indicative of core collapse. Unsurprisingly, these are the clusters with the shortest relaxation times.

Additional long-term internal evolutionary effects

Mass segregation: When a system has a shorter relaxation time, stars will interact very frequently, exchanging energy towards equipartition. So far, we assume all stars have the same mass, leading also to an equipartition of the velocity distribution. However, a more realistic scenario would feature a range of stellar masses. In this case, massive stars will transfer energy, on average, to lower mass stars, leading to mass segregation, where the lighter stars speed up, and the massive ones slow down, sinking to the cluster centre.

Post core-collapse evolution: As density rises during core collapse, the number of star-star interactions increases significantly, even leading to physical collisions – usually deemed unlikely in the densities expected of galaxies. There are typically hundreds of collisions during the lifetime of a dense cluster. Blue stragglers are stars that appear on the Main Sequence above the turn-off point; i.e., they are more massive than expected from the age of the cluster. These stars were created by collisions. Such a high density environment can also lead to the formation of a singularity (i.e., black hole), formed at the centre of the cluster and expected to be of intermediate mass ($\sim 10^3 M_\odot$). A few possible cases are known so far (e.g., M15). The paucity of collapsed cores is likely the result of a mechanism that opposes this process: heating by binary systems, an effect we will see next.

Stellar binaries and Heggie's law: So far, we have considered only individual stars, but they can exist as binary systems, and their internal energy can help 'absorb' the increasing binding energy as the core collapses. Binaries can be formed in a primordial way. Pre-collapsed cores contain stars formed already as binaries. This is still a relatively unknown quantity, but it could be very significant. They could also form via three-body encounters, or even in a two-body encounter if tidal interactions are involved, so that variations in the internal energy can compensate for the evolution from an unbound (positive energy) to a bound system (negative energy). Encounters with binaries will change the orbits of stars, acting as a sink or a source of gravitational energy.

Binaries can be defined as soft or hard depending on whether their binding energy is less or greater than the mean kinetic energy of individual stars, respectively. Dynamical studies show that as a result of encounters, soft binaries ($E_b \ll m\sigma^2$) get softer (decreasing $|E_b|$), and hard binaries ($E_b \gg m\sigma^2$) get harder (increasing $|E_b|$). This is known as 'Heggie's law'.[1] This behaviour can be explained via equipartition. Consider a soft binary. If the kinetic energy of field stars is greater than the kinetic energy of stars in the binary, the internal energy of the binary will be raised, becoming softer. This process will eventually lead to the break up of the binary, operating over timescales shorter than the relaxation timescale. We can envisage an equilibrium state when the rate of binary loss equates the rate of binary formation via three-body encounters. Hence, this process is not important in the evolution of the core. In the opposite case, a hard binary has a higher kinetic energy than a typical field star. The interaction will transfer energy to the field star, making the binary more bound (harder). This can be an important energy source in dense star clusters, and could stop core collapse, and even drive the expansion of the core, at the expense of hardening the binary population. The exchange of energy with binaries makes X-ray binaries (systems that emit in X-ray because of the infall of gas from a main sequence star to a compact object in a close binary) more common in the central regions of globular clusters.

Exercise 8.2

Let us describe a GC as an isothermal sphere, truncated at $r_0 = 5$ pc, with total mass $M = 10^6 M_\odot$. If a mass fraction, f, is in the form of close binaries, consisting of two $1 M_\odot$ stars orbiting with a separation $a = 10 R_\odot$, determine the value of f for which the binding energy in binaries equals the total binding energy of the GC.

8.3 External effects: Tidal disruption

Stellar systems are affected by their motion within the host galaxy. Being extended systems, they are especially prone to tidal interactions. Tidal forces arise from the fact that an extended distribution of mass feels different gravitational forces in different regions (such is the case with the Moon-Earth system). Therefore, stars in the outer regions of a cluster are substantially affected either by the potential of the galaxy or by the passage of nearby masses. Moreover, for a GC with an eccentric orbit outside of the Galactic plane, the passage through the disc produces a tidal pulse. We can therefore classify tidal effects as either continuous processes caused by the motion within the potential of the galaxy, or tidal shocks, where the effect is experimented during a relatively short timescale. We will consider these two mechanisms below.

Steady tidal forces: Jacobi radius

Consider a cluster orbiting its host galaxy. Tidal forces exerted by the gravitational potential of the galaxy will be able to strip stars from the cluster.

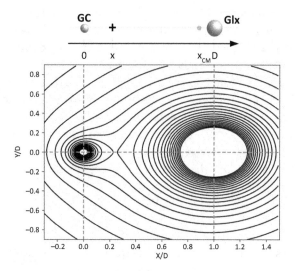

Figure 8.2 Simplified version of the steady tidal interaction as a globular cluster (GC) orbits around the Galaxy (Glx).

Therefore, at a given radius (the tidal radius) the number density of stars will drop abruptly as their orbits are no longer bound to the cluster.

Let us explore a simple case, where the cluster (mass m) orbits the galaxy (mass M), at a distance D, on a circular orbit, beyond the outer edge of the host system. Both galaxy and cluster are spherical. The angular speed of the orbital motion (around the common centre of mass) is

$$\omega = \sqrt{\frac{G(M+m)}{D^3}}. \tag{8.10}$$

The total energy per unit mass of the cluster can be written in terms of the effective potential:

$$\mathcal{E} = \frac{1}{2}v^2 + \Phi_{\text{eff}}(r) = \frac{1}{2}v^2 + \Phi(r) + \frac{J_z^2}{2r^2}. \tag{8.11}$$

This effective potential – which includes the contribution of angular momentum to the energy – introduces a zero-velocity surface, such that orbits will never enter the region where $\mathcal{E} < \Phi_{\text{eff}}(r)$. Figure 8.2 shows contours of the effective potential for the simplest case of two orbiting point masses. The real thing will be roughly similar. At some critical value of the distance, we have the last zero-velocity contour for the cluster. Hence, we should identify the tidal radius at this point ($r = r_J$), such that

$$\left(\frac{\partial \Phi_{\text{eff}}}{\partial r}\right)_{r_J} = 0. \tag{8.12}$$

In our simple model with two point masses we have

$$\frac{GM}{(D - r_J)^2} - \frac{Gm}{r_J^2} - \frac{G(M + m)}{D^3}\left(\frac{D}{1 + m/M} - r_J\right) = 0. \tag{8.13}$$

Given that $m \ll M$, we have $r_J \ll D$, and we can expand $(D - r_J)^{-2}$ as a Taylor series, and truncate to the lowest order:

$$r_J \simeq D\left(\frac{m}{3M}\right)^{1/3}. \tag{8.14}$$

This is the Jacobi limit of the mass m, which provides an estimate of the tidal radius of a cluster. There are several aspects that can make r_J differ from the tidal radius:

+ The zero velocity surface is not spherical (a single radius is not valid).
+ Noncircular (stellar) orbits can be bound to the cluster, even though they probe out to $r > r_J$.
+ The cluster does not move along a circular orbit. The difference will be especially large with highly elongated orbits.
+ The cluster orbits within the host.

In any case, the Jacobi limit from equation 8.14 gives values around tens of parsec, similar to the cutoff radii found in the surface brightness profiles of globular clusters.

Tidal shocks

In these encounters – for example, between open clusters and giant molecular clouds – the stars are given a sudden jolt. On average, the energies will increase. Let us consider the effect on a static star in the cluster located close to the centre, after the passage of a cloud (see figure 8.3). Δv is perpendicular to the direction of motion of the cloud. (This is similar to the relaxation time approximation we did in section 3.3.) Summing up the effect in all stars in the cluster gives the total change in energy:

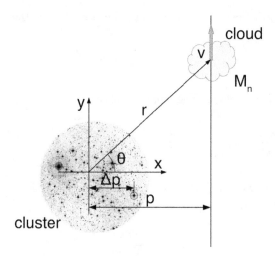

Figure 8.3 Illustration of a tidal shock as a cloud passes near a stellar cluster.

$$\Delta E = \frac{1}{2} M_c (\Delta v)^2. \tag{8.15}$$

We follow by integrating over all encounters with clouds at all impact parameters, to get

$$\frac{dE}{dt} = 4\pi G^2 \frac{M_c R_c^2}{3v} \rho_{\text{neb}} \frac{M_{\text{neb}}}{R_{\text{neb}}^2}. \tag{8.16}$$

We can use the virial theorem $2T + W = 0 \Rightarrow E = W/2$ and take a generic expression of the potential energy as

$$W = -\gamma \frac{GM_c^2}{R_c}, \tag{8.17}$$

where γ is, once more, a factor depending on the geometry of the cluster. Hence

$$\frac{dE}{dt} = \frac{\gamma}{2} \frac{GM_c^2}{R_c^2} \frac{dR_c}{dt}. \tag{8.18}$$

Comparing equations 8.16 and 8.18, we get

$$\frac{dR_c}{dt} = \frac{8\pi G}{3\gamma v} \rho_{\text{neb}} \frac{M_{\text{neb}}}{M_c} \frac{R_c^4}{R_{\text{neb}}^2}. \tag{8.19}$$

We can relate this equation to the evolution of the density in the cluster:

$$\frac{d\rho_c}{dt} = -\frac{6G}{\gamma v}\rho_{\text{neb}}\frac{M_{\text{neb}}}{R_{\text{neb}}^2}, \qquad (8.20)$$

which means the cluster density decreases at a constant rate. The lifetime of the cluster against tidal shocks can be estimated by assuming the change in density is equivalent to the density itself ($\Delta\rho_c \sim \rho_c$), leading to

$$t_{\text{shock}} = \frac{\gamma v}{6G}\frac{\rho_c}{\rho_{\text{neb}}}\frac{R_{\text{neb}}^2}{M_{\text{neb}}}, \qquad (8.21)$$

proportional to ρ_c. Hence, this effect is very important in open clusters (lower density and located on the Galactic plane). Higher density clusters will be less likely to be affected by tidal shocks (rather, affected by cluster evaporation).

Globular clusters would be less affected by tidal shocks because of their higher densities and orbits away from the plane of the disc. However, they can be affected by the passages through the disc, and it is believed that low-mass globular clusters were indeed dissolved during the early phases of the formation of the Galaxy.

8.4 Cluster evaporation: King models

One of the differences between a thermodynamic system and a collisionless set of particles moving under gravity is the issue of escape velocity: In a relaxed cluster, achieving a Maxwellian distribution in the velocities does not imply settling into equilibrium. Stars in the high-velocity tails of the distribution will escape from the gravitational potential. A new equilibrium, after several relaxation timescales, would now imply a slightly lower escape velocity, continuing the evaporation process. The simplest approximation to solve this problem is to follow an iterative method, allowing the cluster to relax to a Maxwellian distribution, then 'chopping off' the high velocity tails at $|v| > v_{\text{esc}}$, etc. This introduces the concept of a loss rate per relaxation period. The distribution function in velocity space is

$$f(v) = \frac{1}{(2\pi)^{3/2}\sigma^3}e^{-v^2/2\sigma^2}, \qquad \text{where } \sigma = \sqrt{\frac{\langle v^2 \rangle}{3}}.$$

From the virial theorem we obtain $\langle v_{\text{esc}}^2 \rangle = 4\langle v^2 \rangle = 12\sigma^2$. So, the fraction of stars moving faster than the escape velocity is

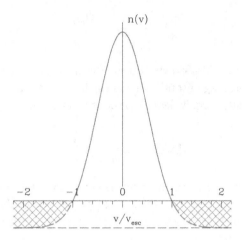

Figure 8.4 Lowered Gaussian distribution to reconcile a Maxwell-Boltzmann profile with the expected ejection of unbound stars from a cluster (hatched region).

$$F = \frac{1}{(2\pi)^{3/2}\sigma^3} \int_{v_{esc}}^{\infty} e^{-v^2/2\sigma^2} 4\pi v^2 dv = \frac{4}{\sqrt{\pi}} \int_{\sqrt{6}}^{\infty} e^{-x^2} x^2 dx = 0.007.$$

A more accurate estimate using the Fokker-Planck equation, assuming all stars have the same mass, gives F = 0.008. So, roughly 1 per cent of the stars escape from the globular cluster during each relaxation time. Therefore we can approximate the distribution function as a lowered Maxwellian velocity distribution (see figure 8.4), where the value of the phase space distribution function is identically zero for speeds greater than v_{esc}. There are two aspects that simplify the dynamical modelling of globular clusters:

+ $t_{relax} \gg t_{crossing} \Rightarrow$ the collisionless Boltzmann equation (CBE) can be used.
+ $t_{relax} \ll$ age of the cluster $\Rightarrow$ clusters are relaxed.

We need to obtain a self-consistent density-potential pair (ρ, Φ) compatible with the lowered Gaussian distribution function. For a spherically symmetric system there is only one spatial variable, r; the velocity can be radial or transverse, (v_r, v_t); and there are only two integrals of motion: Energy $(\mathcal{E} = v^2/2 + \Phi)$ and angular momentum $(J = rv_t)$. Therefore, the density profile can be written:

$$\rho(r) = \int f(r, v_r, v_t) 2\pi v_t dv_t dv_r = 2\pi \int F(E, J) v_t dv_t dv_r, \qquad (8.22)$$

where the second equality derives from Jeans theorem. The other equation needed to relate the density and potential is Poisson's equation (equation 2.17), in spherical coordinates:

$$\frac{1}{r^2}\frac{d}{dr}\left(r^2\frac{d\Phi(r)}{dr}\right) = 4\pi G\rho(r). \tag{8.23}$$

King models provide a practical representation of the density and potential of a globular cluster. They make the following assumptions:

1. The velocity distribution is isotropic, so that $F = F(E)$.
2. Because of external tidal forces, the cluster has a finite radius (i.e., the Jacobi radius), where we define the velocity to be zero.
3. The velocity distribution at $r = 0$ is given by the lowered Gaussian approximation, namely:

$$f(r=0, v) = \begin{cases} k\left(e^{-v^2/2\sigma^2} - e^{-v_{esc}^2/2\sigma^2}\right) & v \le v_{esc}, \\ 0 & v > v_{esc}. \end{cases} \tag{8.24}$$

Notice σ is the root mean square dispersion along any of the components of the velocity. By defining $f(0, v)$, we can calculate $f(r, v)$ at all radii by using $v_{esc}^2 = -2\Phi(r)$. From this we can derive the density distribution:

$$\rho(r) = 2^{5/2}\pi k\sigma^3 e^{\Phi_0/\sigma^2}$$
$$\times \left[\frac{\sqrt{\pi}}{2}e^{-\Phi/\sigma^2}\,\mathrm{erf}\left(\sqrt{\frac{-\Phi}{\sigma^2}}\right) - \sqrt{\frac{-\Phi}{\sigma^2}} - \frac{2}{3}\left(-\frac{\Phi}{\sigma^2}\right)^{3/2}\right], \tag{8.25}$$

and the gravitational potential:

$$\Phi(r) - \Phi_0 = G\int_0^r \frac{M(<s)}{s^2}ds = 4\pi G\left[-\frac{1}{r}\int_0^r \rho(s)s^2 ds + \int_0^r \rho(s)s\,ds\right]. \tag{8.26}$$

The integral is solved by parts ($\int u\,dv = uv - \int v\,du$), choosing $u = M(<s)$ and applying the Leibnitz integral rule. These equations are solved by choosing a central value for the potential (Φ_0) and integrating the above equations outwards until $\Phi(R) = 0$. We need to solve these equations via numerical integration, yielding a family of models characterized by the concentration index:

$$c \equiv \log\left(\frac{r_t}{r_0}\right), \tag{8.27}$$

where $r_t = R$ is the tidal radius (i.e., the 'total' radius of the globular cluster), and r_0 is the King radius, defined by

$$r_0 \equiv \sqrt{\frac{9\sigma^2}{4\pi G \rho_0}}, \tag{8.28}$$

which corresponds to the radial distance where the projected density of the isothermal sphere is 0.5013 of its central value; hence r_0 is close to the definition of the core radius. Typically, the concentration lies in the region $0.5 < c < 2.3$. King models give acceptable fits to globular and open clusters. They can also be applied to model the surface brightness profile of dwarf spheroidal and elliptical galaxies. With a knowledge of r_0 and σ^2 (observed), we can infer the central density, hence obtaining the central mass-to-light ratio. This technique is known as 'core fitting'. Most clusters are therefore modelled by three parameters: r_c, r_t, ρ_0. In the central region, the projected two-dimensional surface mass density (see section 2.6) can be approximated by

$$\Sigma(r) = \frac{\Sigma_0}{1 + \left(\frac{r}{r_c}\right)^2}, \tag{8.29}$$

which can be extended to a surface brightness profile if we make an assumption about $\Upsilon = M/L$ (usually kept constant). This function gives a good aproximation to the observed profiles of elliptical galaxies. The three-dimensional equivalent is the King profile presented in chapter 2.

For a galaxy (or an unrelaxed cluster), the assumption of isotropy may not be well justified, and we have to consider anisotropic velocity distributions. For instance, the orbits in the outer parts of the cluster may be mainly radial.

Note

1 Heggie, 1975, MNRAS, 173, 729.

9
Larger stellar systems: Galaxy clusters

Beyond galaxy scales, an extended structure pervades the Universe in the form of filaments over very large distances ($\gtrsim$10s of Mpc). At the confluence of these filaments lie clusters of galaxies. We will briefly explore their properties in this final chapter, in relation to all the material seen so far. A simple model including hydrostatic equilibrium allows us to measure the masses of galaxy clusters, leading to yet another robust proof of the presence of dark matter. In addition, gravitational lensing based on the deflection of light by the presence of large masses is a useful technique. We will present the lensing equation and its application to simple mass distributions, showing how one can measure cluster (and galaxy) masses. The use of galaxy clusters as cosmological probes is also explored here, along with the imprint of large-scale structure on galaxy formation.

9.1 The most massive structures

Galaxy clusters are gravitationally bound structures comprising several hundred galaxies extended over a radius $\gtrsim$1 Mpc. Measurements of galaxy redshifts towards a cluster reveal an overdensity at the cluster redshift, with a velocity dispersion around $\sim$700–1,000 km s^{-1}. Clusters shine brightly at X-ray wavelengths (figure 9.1). The X-ray light is produced by a large amount of diffuse gas at very high temperatures (kT$\sim$keV). The morphology of the galaxy distribution and the X-ray emission is varied, sometimes featuring substructure and irregular shapes, but often showing the smooth morphology expected of a system in equilibrium. Clusters are the largest virialized structures in the Universe. In the bottom-up hierarchical formation scenario, massive clusters assemble at later times (see section 7.3). Therefore, in contrast with galaxies, their properties are closer to the expectations from linear theory, making

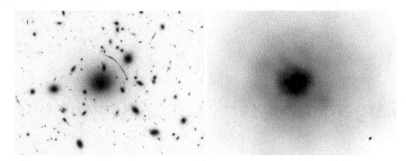

Figure 9.1 Image of galaxy cluster Abell 383 in the optical (*left*, Hubble Space Telescope) and X-ray spectral window (*right*, Chandra Space Telescope). Optical photons originate mostly from the galaxies within the cluster, including a few background galaxies distorted in the shape of arcs by gravitational lensing (see section 9.3). The X-ray image shows the hot diffuse intracluster medium (see section 9.2). (Sources: Optical courtesy: NASA/STScI; X-ray courtesy: NASA/CXC/Cinestav/T.Bernal et al.)

cluster samples valuable cosmological probes. The virial theorem (see section 3.9) allows us to determine the typical mass range of a galaxy cluster. For an isothermal density profile, we find that

$$M \sim \frac{3\sigma^2 R}{G} \approx 6.9 \times 10^{14} M_\odot \left(\frac{\sigma}{1,000\,\mathrm{km\,s^{-1}}} \right)^2 \left(\frac{R}{1\,\mathrm{Mpc}} \right). \quad (9.1)$$

Making the simple assumption that a rich cluster includes the equivalent of 100 galaxies with an average luminosity similar to the Milky Way ($4 \times 10^{10} L_\odot$) gives a mass-to-light ratio $\Upsilon \sim 200 \Upsilon_\odot$, significantly higher than any stellar population would produce. Hence the contribution from stars to the total mass budget in clusters is negligible. In the next section we will see that X-ray observations allow us to constrain the gas mass as well, which, although overwhelmingly higher than the stellar mass content, is still short of the masses needed to explain the high velocity dispersion. Clusters are strongly dark matter–dominated systems (see figure 1.6), and produce very interesting observational constraints about the interplay between dark matter and baryons.

9.2 X-ray measurements of the cluster mass

An important tracer of galaxy clusters is the gaseous component pervading the cluster in a hot diffuse form. It includes the vast majority of the baryon matter content, up to 80–90 per cent by mass. The presence of this so-called intra-cluster medium (ICM) can be explained following the cooling argument invoked in section 7.6. The low density of the ICM results in very long cooling timescales, preventing the collapse and eventual transformation of this gas into stars. It can be described by a relatively simple model, following the equation of state of an ideal gas:

$$p = nkT = \left(\frac{\rho_g}{\mu m_P}\right) kT, \qquad (9.2)$$

where μm_P is the mean weight of a gas particle, and m_P is the proton mass. The gas is assumed to be in hydrostatic equilibrium within the gravitational potential of the underlying dark matter halo. This model allows us to derive the mass of a galaxy cluster by observations in the X-ray spectral window. We obtain

$$\frac{dp}{dr} = \frac{kT}{\mu m_P}\frac{d\rho_g}{dr} + \frac{\rho_g k}{\mu m_P}\frac{dT}{dr} = -\frac{GM(<r)}{r^2}\rho_g, \qquad (9.3)$$

leading to the following expression:

$$M(<r) = -\frac{kTr}{G\mu m_P}\left(\frac{d\ln T}{d\ln r} + \frac{d\ln\rho_g}{d\ln r}\right). \qquad (9.4)$$

How hot is this gas? By use of the virial theorem, and relating the kinetic energy per particle to the thermal energy, we get

$$\frac{3kT}{\mu m_P} \sim \frac{GM}{R} = GM\left(\frac{4\pi\rho}{3M}\right)^{1/4}. \qquad (9.5)$$

The density of the cluster can be obtained from the spherical collapse model, adopting the Planck 2015 cosmological parameters (see table 7.1), namely:

$$kT = 0.39\,\text{keV}\,M_{14}^{2/3}\mu_{0.5}(1 + z_{\text{VIR}}), \qquad (9.6)$$

where M_{14} is the cluster mass in units of $10^{14}M_\odot$, and $\mu_{0.5} = \mu/0.5$. Note the virialization of a cluster takes place at relatively low redshift $z_{\text{VIR}} \sim 1$. Therefore, typical cluster masses yield ICM temperatures in the range ~ 1 keV, i.e., in the X-ray spectral window. The dominant emission of the hot ICM gas is produced by thermal bremsstrahlung,[1] so that the emissivity at frequency ν is

$$J(\nu) \propto \rho_g^2 T^{-1/2} e^{-h\nu/kT}. \qquad (9.7)$$

A spatially resolved observation of the surface brightness profile in X-ray ($\propto J$), including its spectral shape ($\propto dJ/d\nu$), allows us to determine the density and temperature profile of the gas, which, via equation 9.4, leads to the total mass profile M($<$r). A density distribution commonly used to model the dark matter in clusters is the King profile (see equation 2.34). Noting the gas is in hydrostatic equilibrium (see exercise 3.6), we understand that the gas density is described by the so-called β-model:

$$\rho_g(r) = \rho_0 \left[1 + \left(\frac{r}{r_c} \right)^2 \right]^{-3\beta/2} , \qquad (9.8)$$

where $\beta = \mu m_p \sigma^2 / kT$ is the ratio between the thermal energy of the gas and the typical energy of a dark matter particle. By fitting this model to the observations, we can derive the gas and total mass profile. A comparison of the total mass with the luminosity in the optical window, i.e., contributed by galaxies, gives values of the mass to light ratio of $\Upsilon \gtrsim 200 \Upsilon_\odot$, compatible with our previous virial mass estimate (equation 9.1). Moreover, integrating the gas density profile (equation 9.8) with the constraints from X-ray emission produces a mass content that amounts to over 80 per cent of the total *baryon* budget in clusters. Note that even in this case, the total baryon mass to light ratio would be $\Upsilon_b \sim 50 \Upsilon_\odot$, still lower that the expectation from the virial theorem, and unmistakably requiring a large fraction of the cluster mass in the form of dark matter.

9.3 Gravitational lensing

The gravitational field of a mass distribution distorts spacetime, bending the path of photons. This effect is especially strong when the photons originate from a distant background source along a very similar line of sight to the observer. We can derive the lensing equation by adopting the thin lens approximation, a simple, but realistic assumption that the extent of the perturber (i.e., the lens) is much smaller than the distance between the lens and the source (D_{LS}) or between the observer and the lens (D_{OL}).

The layout of the lensing problem is shown in figure 9.2. The observer (O) looks at a distant galaxy (S), located very close to a foreground mass distribution that acts as a lens (L). Our measurements are based on angular separations on the sky. The angle θ is the *observed* position of S, at point I on the source plane, whereas the true position – which would have been the observed position, had the lens not been there – is β. The

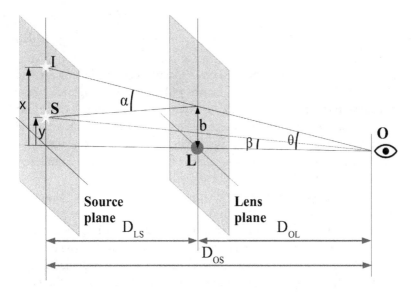

Figure 9.2 Gravitational lensing effect. Light from a background source (S) is deflected by a mass (L) between the source and the observer (O). The deflection angle α causes the observer to image the source at position I. The different parameters shown here help us derive the lensing equation (9.10).

deflection angle caused by the effect of the lens is α. For a point mass acting as lens, the deflection angle of a light ray hitting the lens plane at impact parameter b is

$$\alpha = \frac{4GM}{c^2 b}.\tag{9.9}$$

We assume that the angles considered here are very small ($\theta, \alpha, \beta \ll 1$), so that $\beta = y/D_{OS}$ and $\theta = x/D_{OS} = b/d_{OL}$ and $x - y = \alpha D_{LS}$. The lensing equation (relating observed and true position of the source) is therefore

$$\theta - \beta = \alpha \frac{D_{LS}}{D_{OS}} = \frac{4GM}{c^2} \frac{D_{LS}}{D_{OL}D_{OS}} \frac{1}{\theta} \equiv \frac{\theta_E^2}{\theta},\tag{9.10}$$

where

$$\theta_E \equiv \sqrt{\frac{4GM}{c^2} \frac{D_{LS}}{D_{OL}D_{OS}}}\tag{9.11}$$

is the Einstein radius of the lens. This is a second order equation in the observed position, with solutions

$$\theta^2 - \beta\theta - \theta_E^2 = 0 \Rightarrow \theta_\pm = \frac{\beta \pm \sqrt{\beta^2 + 4\theta_E^2}}{2}. \tag{9.12}$$

If $\beta = 0$ – i.e., the background galaxy is exactly positioned along the line of sight to the lensing mass – the source will appear as a ring with radius θ_E (Einstein ring). Otherwise, the equation produces two solutions: one image 'inside the ring' ($\theta < \theta_E$) and another one 'outside'. For an extended mass distribution with spherical symmetry, the solution is analogous, replacing the mass in the deflection angle (equation 9.9) by the cumulative mass contained within the impact parameter, $M(< b)$.

Figure 9.3 shows an example of the use of the image positions to solve for the mass contained inside. Note that some asssumption has to be made regarding the mass density profile of the lens. For a case close to total alignment ($\beta \ll 1$), the mass contained within the Einstein radius should be independent of this assumption. In general, lensing mass estimates tend to have uncertainties that sharply decrease at the equivalent of the Einstein radius. As the true position of the source departs outwards ($x \equiv \theta_E/\beta \ll 1$), one of the solutions is very close to the true position, $\theta_+ = \beta(1 + 2x)$, and the other one is close to the origin ($\theta_- = -2\beta x$) and is therefore not visible as it is located at the position of the lens. This behaviour defines the weak lensing regime, whereas $\beta \sim \theta_E$ corresponds to strong lensing. If we relax the assumption of spherical symmetry for the lens mass distribution, more complex solutions are produced.[2] To put in context the expected behaviour in galaxies and clusters, the deflection angle in typical units of these systems are

$$\alpha = 0.4\,\text{arcsec}\left(\frac{M}{10^{11}M_\odot}\right)\left(\frac{b}{10\,\text{kpc}}\right)^{-1}$$
$$= 40\,\text{arcsec}\left(\frac{M}{10^{15}M_\odot}\right)\left(\frac{b}{1\,\text{Mpc}}\right)^{-1}, \tag{9.13}$$

making strong lensing events available over both galaxy and cluster scales – noting that typical ground-based images without adaptive optics can reach a spatial resolution ~ 0.5-1.0 arcsec, and the Hubble Space Telescope features a typical resolution ~ 0.1 arcsec. An additional quantity often used in gravitational lensing is the critical surface mass density:

$$\Sigma_c \equiv \frac{c^2 D_{OS}}{4\pi G D_{LS} D_{OL}}. \tag{9.14}$$

This expression depends only on the geometry of the lens, i.e., the angular diametre distance to lens and source and between them, given by the

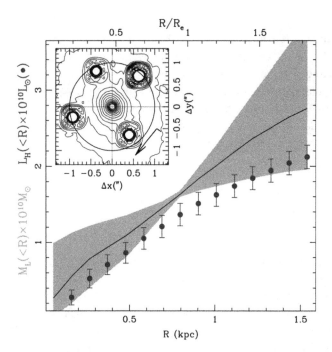

Figure 9.3 Luminosity (dots) and lensing mass profile (shaded area) of the Einstein Cross, a lensing system that produces four images of a background quasar. Note that the uncertainties in the lensing mass produce a typical "butterfly" diagram, where the uncertainty is lowest at the position of the Einstein radius, represented by a circle in the inset. (Source: Ferreras et al., 2010, MNRAS, 409, L30.) Used by permission of Oxford University Press.

redshifts and the adopted cosmology. Σ_c represents the typical surface mass density needed for the presence of a detectable lensing signal, i.e., with large enough θ_E. This is especially evident when comparing it with the definition of the Einstein radius:

$$\Sigma_c = \frac{M}{\pi (\theta_E D_{OL})^2}. \tag{9.15}$$

Exercise 9.1

Consider the strong gravitational lens HE1104-1805[a], a galaxy at redshift $z_L = 0.73$ that produces two images of a background

---

source at redshift $z_S = 2.32$, with an apparent angular separation of 1.1 and 2.1 arcsec away from the galaxy in opposite directions. Taking a fiducial ΛCDM cosmology ($h = 0.7$; $\Omega_{m,0} = 0.3$), the angular diameter distances are: $D_{OL} = 1498$ Mpc; $D_{OS} = 1690$ Mpc; and $D_{LS} = 910$ Mpc. Assuming a point source model for the lens, calculate the lensing mass. What would be your answer if you adopt an isothermal profile?

---

[a]https://www.cfa.harvard.edu/castles.

9.4 Clusters and cosmology

In chapter 7 we traced the evolution of galaxies from the linear regime, where the density fluctuations are very small, to the collapse and virialization of a structure. We could also appreciate how the standard ΛCDM (see chapter 7) framework leads to a bottom-up formation, so that the most massive structures virialize, on average, at later times. Galaxies are supposed to collapse at rather early cosmic times with respect to the present, so that linear theory is not applicable to describe the properties of nearby galaxies. In contrast, clusters, or, even better, superclusters, are more massive structures whose formation took place at later times, thereby allowing us to use linear theory to describe their evolution. The distribution of clusters and their evolution with redshift thus provides a useful probe of the cosmology. For instance, one of the key cosmological parameters that describes the amplitude of the power spectrum of density fluctuations is σ_8, i.e., the variance (when squared) of the mass fluctuations over a scale of $8h^{-1}$ Mpc, a size that engulfs the typical mass of a galaxy cluster. One of the most direct applications of the Press-Schechter methodology (see section 7.4) is to infer this normalization from the number density of clusters at present time. Observations of X-ray clusters at low redshift[3] give a number density of $M_c = 10^{15} M_\odot$ clusters around 3.8×10^{-8} Mpc^{-3} within a mass interval of $10^{14} M_\odot$ (in this exercise we are assuming $h = 0.7$ and $\Omega_{m,0} = 0.3$). We use equation 7.61, noting that the average background density at present time is $\langle \rho \rangle = \rho_{\mathrm{crit},0} \Omega_{m,0} = 4.1 \times 10^{10} M_\odot$ Mpc^{-3}. The mass variance $\sigma(M)$ can be related to the radius (R) of the spherical top-hat filter over which we perform the Press-Schechter smoothing:[4] $\sigma(R) = \sigma_8 R_8^{-\gamma}$, where R_8 is the radius in units of $8h^{-1}$ Mpc, and $\gamma = 0.8$.

Finally, we get

$$n(M_c) = 3.8 \times 10^{-8} \frac{1}{\mathrm{Mpc}^3 10^{14} \mathrm{M}_\odot} = 1.2 \times 10^{-6} \frac{1}{\mathrm{Mpc}^3 10^{14} \mathrm{M}_\odot} x e^{-x^2},$$

where $x \equiv \delta_c R_8^\gamma / \sigma_8 \sqrt{2}$. The solution to this equation is $x \approx 2.04$. If we consider that the linear density contrast for collapse is $\delta_c \sim 1.7$ (see equation 7.57) and take $R_8 = 1$, we infer a normalization of $\sigma_8 = 0.6$. This very simple exercise gives the right ballpark value (see table 7.1), illustrating how clusters can be approximately considered 'linear structures' to characterize the normalization of the power spectrum.[5]

9.5 Environment-related processes

In addition to the cosmological importance of galaxy clusters as tracers of structure growth at late times, they also enable us to probe the various mechanisms of the baryon physics that drives galaxy formation. The simple description of an isolated 'structure unit' comprising a galaxy living in the centre of a virialized dark matter halo, along with an additional diffuse gas component pervading the halo, breaks down when we take into account the complex environment of galaxies over larger scales, with such units under constant gravitational interaction within a highly nontrivial cosmic web (see figure 1.4). In this context, clusters represent regions where galaxies experiment the extreme case of environment-related processes. Large surveys in combination with highly detailed numerical models of galaxy formation in a cosmological context have allowed us to visualize the potential factors that regulate the evolution of galaxies. The most important environment-related effects, in no particular order, are:

1. Ram-pressure stripping, suffered by the gas component in galaxies as they fall into a cluster: This process will be more efficient at increasing infall velocity and at lower surface gas density (i.e., the outer regions of galaxies are more prone to stripping). The pressure is exerted by the hot intracluster medium. This process is one of the factors that control the quenching of star formation in galaxies as they enter clusters, potentially transforming spiral galaxies into lenticulars (S0).

2. Harassment, which is a repetitive process of gravitational interaction among galaxies at high speed: Like the tidal shock scenario presented in section 8.3, each high-speed passage acts as an impulsive force on the stars and gas, increasing the energy of the galaxy, whose components

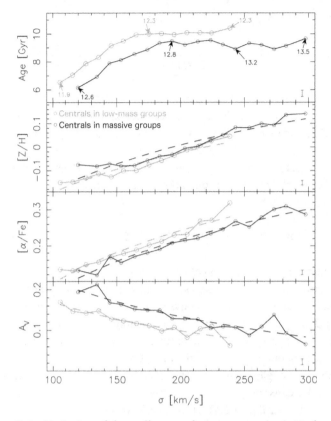

Figure 9.4 Variation of the stellar population properties in Early-type galaxies according to their local 'mass' (parameterized by the velocity dispersion, σ) and their environment (split into galaxies in low- and high-mass groups). From top to bottom, the age, metallicity, abundance ratio and dust attenuation are shown with respect to σ. At fixed galaxy mass, galaxies in more massive halos appear younger, dustier and with a lower [α/Fe], representative of a more extended star formation history. (Source: La Barbera et al., 2014, MNRAS, 445, 1977.) Used by permission of Oxford University Press.

become less bound. In a cluster environment, these events will be quite repetitive, leading to the disruption of the galaxies, especially by removing the outer, less bound envelopes, and by triggering dynamical instabilities that will affect the morphology.

3. Strangulation: Star formation in a galaxy can be fuelled by the infall of gas from a diffuse component throughout the halo. As a galaxy falls into a cluster, this reservoir may be depleted. Note the difference from ram

pressure stripping, which removes the interstellar medium of the galaxy, therefore instantaneously truncating the star formation process. The effect of strangulation on the star formation rate is more gradual, extending over larger timescales.

4. Mergers and cannibalism: The interaction between two galaxies and their host halos may lead to a merger, greatly helped by dynamical friction (see section 3.10). Note that a merger process is more effective if the relative velocities of the galaxies is comparable to the velocity dispersion of the underlying stellar components. Therefore, the cores of massive clusters, with relative velocities $\sim 1,000$ km s^{-1} are not fertile ground, but lower mass groups and the outskirts of clusters are regions where galaxy mergers may be present. An additional merging process, termed 'cannibalism', can be quite efficient in the central regions of clusters, where galaxies will lose energy via dynamical friction, merging with the central, most massive galaxy.

Therefore, if we assume that, to the lowest order, the formation of a galaxy is driven by its total mass (i.e., a *local* property), the combination of all these environment-related processes produces an entangled mixture of star formation histories that greatly complicate the analysis and extend the formation to a *global* context. In some cases, for instance choosing early-type galaxies – whose more homogeneous stellar populations are easier to study – it is possible to quantify the role of environment, as shown in figure 9.4, where the spectra of early-type galaxies from the Sloan Digital Sky Survey were used to constrain the properties of the stellar populations. At fixed velocity dispersion (a rough proxy of galaxy mass), central galaxies living in more massive halos appear slightly younger, with lower [α/Fe], reflecting a more extended star formation history than their counterparts in lower density environments. This effect can be explained by the contribution of infalling satellite galaxies.

Notes

1 Bradt, 2008, *Astrophysics processes*, Cambridge, section 5.5.
2 Schneider, Kochanek & Wambsganss, 2006, *Gravitational lensing: Strong, weak and micro*, Springer.
3 E.g., Reiprich & Böhringer, 2002, ApJ, 567, 716.
4 Viana & Liddle, 1996, MNRAS, 281, 323.
5 Allen, Evrard & Mantz, 2011, ARA&A, 49, 409.

Further reading

Arfken, George B., Hans Weber and Frank E. Harris. *Mathematical Methods for Physicists*. 7th ed. Amsterdam: Academic Press, 2013. A great reference for those keen on strengthening their maths/calculus background, especially aimed at physics students.

Bertin, Guiseppe. *Dynamics of Galaxies*. Cambridge: Cambridge University Press, 2014. An insightful presentation of galaxy dynamics, with a very interesting connection to the properties of plasma, another physical system that behaves as a collisionless fluid.

Binney, James and Scott Tremaine. *Galactic Dynamics*. Princeton, NJ: Princeton University Press, 2008. One of the classic references to the topic, used by many researchers over the years, which should become 'bedtime' reading for any advanced student with a strong interest in pursuing a career in galaxy dynamics.

Carroll, Bradley W. and Dale A. Ostlie. *An Introduction to Modern Astrophysics*. 2nd ed. Cambridge: Cambridge University Press, 2017. An introductory level reference for undergraduate students interested in finding out more about general aspects of astrophysics.

Coles, Peter and Francesco Lucchin. *Cosmology: The Origin and Evolution of Cosmic Structure*. 2nd ed. Chichester: Wiley-Blackwell, 2002. A comprehensive reference to cosmology, with a detailed treatment of the standard cosmological model, structure growth, and observational tests.

Jones, Mark H., Robert J. A. Lambourne and Stephen Serjeant. *An Introduction to Galaxies and Cosmology*. Cambridge: Cambridge University Press/Open University, 2015. General introduction to the wide discipline of cosmology and galaxy formation, which should be a good reference to students without a background in this field.

Longair, Malcolm S. *Galaxy Formation*. 2nd ed. Astronomy & Astrophysics series. Berlin: Springer, 2007. An interesting review of the astrophysics of galaxy formation, with a profusion of references pointing to key research papers in the field, linking the work presented in this book to actual research (although the references may get outdated in this highly active area).

Mo, Houjun, Frank van den Bosch and Simon White. *Galaxy Formation and Evolution*. Cambridge: Cambridge University Press, 2010. A massive tome with "everything you need to know" about galaxy formation and evolution.

Pagel, Bernard. *Nucleosynthesis and Chemical Evolution of Galaxies*. 2nd ed. Cambridge: Cambridge University Press, 2009. A focused treatment of galactic chemical enrichment, written by one of the pioneers in the field.

Sparke, Linda S. and Jay S. Gallagher. *Galaxies in the Universe: An Introduction*. 2nd ed. Cambridge: Cambridge University Press, 2011. Highly pedagogical textbook with many interesting and easy-to-follow problems, useful especially for those with a background in general physics.

Index

Italic numbers denote reference to illustrations; bold numbers denote tables.

Abell 383 galaxy cluster, *164*
Abel transform, 41
abundance matching, 18
abundance ratio [α/Fe], 107, *119*, 119–20, *172*
Active Galactic Nucleus (AGN), 2, 11, 18, 19, 108
alt-azimuthal coordinates, 72–3
anisotropy parameter (β), 59, 63
asymmetric drift, 88–9

bar in spiral galaxies, 6, 20, 71, 104, *105*
baryonic matter
  density fluctuations of, 133–4
  and galaxy formation, 1, 13–14, 93
  oscillations of, 135, 144, 145
baryon physics, 137
beta model, 166
Big Bang, 107, 133
binaries, 154–5
black hole, 3, 66, 110, 154. *See also*
     supermassive black hole (SMBH)
blue stragglers, 154
bolometric flux, 23
BPT diagram, 4–5
bulges, 7, 20, 71, 93
bulge-to-disc ratio, 93

Cartesian coordinate system, 73
celestial sphere, 72
Cepheids, 12
chemical elements
  C, 107
  Fe, 107, 108–9, 118–20

H, 108–9
Mg, 107
O, 107
synthesis of, 21, 107, 108, 112
as tracers of galactic history, 20–1, 107, 108, 118–20, 173
cluster evaporation, 159
cold dark matter (CDM), 14, 127, 135, *144*, 145
collision, definition of, 51
collisionless Boltzmann equation (CBE), 43, 52–3, 60, 61–2, 85–7
colour index
  definition of, 2
  information provided by, 24
  as proxy of stellar age, 2, 3
coordinate systems of celestial objects, 72–3
corotation, 103, 105
correlation function, 142
Cosmic Microwave Background (CMB), 121, 122, 133
cosmological constant (Λ), 124, 125
cosmological distance ladder, 12
cosmological models, *144*, 144–5
cosmological parameters, 145–6, **146**
cosmology, 15, 121, 122, 128–9, 170
Coulomb logarithm, 47, 51, 68
critical density (cosmology), 125
critical surface mass density (gravitational lensing), 168

dark matter. *See* cold dark matter (CDM); hot dark matter (HDM); warm dark matter (WDM)

dark matter (*cont.*)
  density fluctuations of, 133–4, *134*, 135
  distribution of, 1
  in formation and evolution of galaxies, role of, 144–5
  fractional contribution by mass, 1, 14
  in galaxy clusters, 164
  types of, 145–6
dark matter halos
  components of, 63
  density and epoch of collapse, 136–7
  distribution of gas in, 146
  formation of, 13, 107, 138, 139–40
  interaction between, 19
  Navarro-Frenk-White profile of, 139
  ratio between stellar mass and mass of, *18*, 18–19
  statistical derivation of mass function of, 138–9, 140
density
  contrast, 14, 130
  critical, 121, 125
  dimensionless, 125–6
density fluctuations
  correlation functions of, 141–4
  description of, 129–30
  formation of halos and, 139–40
  galaxy formation and, 121, 122
  intensity of, 142
  Jean mass and, 131–3
  linear collapse model of, 138–9
  linear growth of, 129–31
  Newtonian treatment of, 124, 129
  spatial distribution of, 131
  spherical collapse model of, 135–8, *136*, 139
density parameters, 125, 126, *126*
density wave theory, 104, 105
differential rotation, 74
diffusion coefficients, 66
disc galaxies
  elliptical orbit of stars in, 103
  formation of, 93
  rotation velocity profile of, 89–90, *90*
  spiral arms in, 7, 100–6
  star distribution in, 6, 103
  star formation rate in, 101
  Tully-Fisher relation in, 97–8
distance in cosmology, 128–9
distance modulus, 23
distribution functions
  definition of, 44–5
  examples of, 55
  isothermal sphere, 56–7
  Mestel disc, 57–8
  Osipkov-Merritt models, 58–9
  in phase space, behaviour of, 53
  polytrope model, 55–6

in velocity space, 159
dynamical friction, 66–9, 173

early-type galaxies
  abundance ratio of, 119, *172*
  age of, *172*
  colour-magnitude relation in, 98–9
  Fundamental Plane of, 96–8
  properties of, 94
  rotational support of, *101*
  rotation *vs.* pressure in, 99–100
  stellar population properties of, *172*
ecliptic coordinates, 73
effective potential, 30–1, 156
Einstein Cross, *169*
Einstein radius, 168, 169
Einstein-de Sitter model, 125–7
elliptical galaxies
  anisotropic velocity dispersion of, 100
  flattening of, 99–100
  formation of, 65, 94
  properties of, 94
environment, galactic, 10, *11*, 171, 172, 173
epicyclic frequency, 80
epicyclic motion of stars, 70, 78, *79*, *81*, 83, 103
equatorial coordinates, 73
ERO galaxies (Extremely Red Objects), 2

Faber-Jackson relation, 97–8
flux density, 22, 23, 24
Fokker-Planck equation, 65–6
Friedmann-Lemaître-Robertson-Walker metric (FLRW), 122, 123–4
Friedmann's equations, 123–5
Fundamental Plane (FP), *96*, 96–7

galactic cannibalism, 173
galactic chemical enrichment (GCE)
  basic equations of, 112–13
  closed box model of, 114–15
  effect of initial mass function on, 109
  infall model of, 115, 117
  instantaneous recycling approximation equation of, 113–14
  leaky box model of, 117
  open box model of, 115–17
  outflow model of, 116–17
  star formation rate as function of, 111–12
  stellar yield component of, 108, 111
galactic coordinates, 73
galactic harassment, 171–2
galactic ram-pressure stripping, 171
galactic rotation, 75, 76, 78–9, 80–1
galactic strangulation, 172–3
galaxy clusters
  baryon matter content of, 164
  cosmological importance of, 170–1
  density of, 165, 166

environment-related effects, 171
gaseous component of, 164–5, 166
gravitational lensing effect, 166–70
mass measurement, 163, 164–6
optical and X-ray images of, *164*
properties of, 163–4
Ram-pressure stripping effect, 171
X-ray emission of, 163–4, 165–6
galaxy evolution
  on colour-mass diagram, *17*
  dynamical timescale of, 20
  environment-related processes and, 10
  feedback mechanisms in, 17–19
  internal dynamical effects, 19–20
galaxy formation
  Active Galactic Nucleus' effect on, 11
  baryonic material in, 137
  cooling timescale of, 146–7
  dynamical timescale of, 146–7
  environment-related processes and, 10, 173
  function of dark matter halos in, *18*, 18–19
  initial conditions of, 13–14
  merging processes in, 16, 19, 94
  nucleosynthesis and, 107–8
  overview of, 1
  parameters controlling, 147–8
  physical mechanisms of, 13–21
  Schechter function, 8–9
  supermassive black hole and, 2
galaxy(-ies)
  age of, 4, 98
  chemical composition of, 4, 20–1
  classification of, 6, 92, 100–1
  cold dynamical systems, 93, 94
  colour-magnitude relation in, 98–9
  components of, 1–2
  distance and angular extent of, 128
  distribution of stars in, 45–6
  dynamical states of, 92–4
  environment of, 10, *172*
  evolution of stellar populations in, 21
  flattening of, 99–100
  gravitational interactions between, 171–3
  hot dynamical systems, 93, 94, 100
  luminosity of, 6, 8–9
  mass distribution in, 1, 90–1, 97, 107, *118*
  metallicity of, 98, *118*
  morphology of, 6–7
  motion of, 92, 99
  nuclear activity of, 11
  properties of, 94–5
  redshift, 12–13, 119, 122
  rotation of, 89–90, 99–100
  scaling relations in, 94–9
  star formation history of, 108
  surface brightness of, 5–6, 92, 95–6, 129
  See also individual types of galaxies
galaxy mergers, 16, 19, 94, 173
galaxy spectra, 3–4, *4*, 5, 6

Gaussian distribution function, *160*
G-dwarf problem, 107, 114, 115, *116*
general relativity, 123, 124
globular clusters (GCs)
  density distribution in, 161
  dynamic modelling of, 160
  effect of tidal shock on, 159
  gravitational potential of, 161
  King models for, 161–2
  in Milky Way Galaxy, 150
  properties of, 22, 149–50
gravitation, Newtonian theory of, 27–8
gravitational constant (G), 28
gravitational field equation, 33
gravitational fluctuations, 14–15
gravitational focusing, concept of, 50
gravitational force, 27–8, 32
gravitational lensing effect, 166–70, *167*, *169*
gravitational potential, 30–1, *31*, 32, 33, 34–5
gravitational potential energy, 35–6, 65, 99
gravitational potential energy tensor, 64
gravothermal catastrophe, 152

Harrison-Zeldovich spectrum, 144
HE1104-1805 lens galaxy, 169–70
Heggie's law, 154–5
Hernquist profile, 39, 42
hot dark matter (HDM), *144*, 145
Hubble constant/parameter, 123
Hubble-de Vaucouleurs tuning fork diagram, 6, *7*, 93
Hubble sequence, 93
hydrostatic equilibrium, 56, 165

initial mass function (IMF), 16, 109–10, 111
instantaneous recycling approximation (IRA), 113
intra-cluster medium (ICM), 165
irregular galaxies, 7
isolating (third) integral, 54, *54*, 70
isothermal sphere, 37–8, 56–7

Jacobi radius, 155
Jaffe profile, 38
Jeans equations
  anisotropy parameter, 63
  applications to Milky Way galaxy, 87–9
  in case of asymmetric drift, 88–9
  derivation of, 60–1
  first order moment, 60–1
  in spherical coordinates, 61–3
  vertical motion and, 88
  virial theorem and, 64–5
  zeroth order moment, 60
Jeans mass, 132
Jeans theorem, 53–4

Kennicutt law, 16, 17, 111–12
Kepler's problem
    density-potential pair for, 37
    formulation of, 29–30
    gravitational potential, 30–1
    phase space of closed orbits in, 44
    total energy in, 59
Kepler's three laws of orbital motion,
    29–31, 31
King profile, 40
Kormendy relation, 98

Lagrangian derivative, 52
late-type galaxies, 74, 93
lensing equations, 167–8
lensing mass profile, 169
lenticular galaxies, 6, 7, 93
Lindblad resonances, 103–5, 104, 105
Liouville's theorem, 53
local standard of rest (LSR), 74
luminosity, 95, 95–6
luminosity distance, 5, 128
luminosity function, 8–9

M51 galaxy, 104, 104
M81 galaxy, 2–3, 3
Magellanic Clouds, 22
magnitude scale, 22
mass-density distribution
    in elliptical galaxies, 39
    projection into two-dimensional surface,
        41, 41–2
mass segregation, 153
Maxwell-Boltzmann distribution, 45–6, 56
Mestel disc, 57–8
metallicity (Z)
    analytic expressions of, 114
    correlation between stellar mass and, 118
    definition of, 108, 118
    of early-type galaxies, 172
    of nearby stars, comparison of, 116
    [Mg/Fe] as a cosmic clock, 118
Milky Way galaxy
    application of collisionless Boltzmann
        equation to, 85–7
    bulge/bar of, 71
    circular motion of stars in, 75
    components of, 70–2
    dark matter halo of, 71–2
    differential rotation in, 74–82
    epicyclic motion of stars in, 82–3
    formation of, 21
    Gaia/DR2 view of, 72
    general description of, 70–3
    Jeans equations' application to, 87–9
    mass distribution models, 90–1
    Miyamoto-Nagai model of, 91, 91
    Oort's constants, 76–7
    open clusters in, 151

rotational motion of stars in, 75–7
stellar halo of, 71
thick disc of, 71
thin disc of, 6, 71
vertical motion of stars in, 83–5, 86–7
Milne model, 131
Miyamoto-Nagai model, 91
moments of velocity, 43, 53, 60, 66, 67

Navarro-Frenk-White (NFW) profile, 38–9
Newton's theorems, 33–4, 34, 35

observables
    colours, 2–3, 24
    distance, 12, 128
    environment, 10, 11
    flux density, 22–3
    luminosity function, 8–9
    morphology, 6–7
    nuclear activity, 11
    redshift, 12–13
    size of galaxy, 8
    spectroscopy, 3–5, 24, 25
    star formation rate, 9–10
    surface brightness, 5–6
Oort's constants, 76–7, 79, 102
open clusters (OCs)
    in Milky Way galaxy, 151
    properties of, 22, 149, 151
orbital speed, 28
Osipkov-Merritt models, 58–9

parsec, 12
Petrosian radius, 8
phase space, 43–4
photometric passbands, 23–4, **24**
photometry, 23, 24–5
Plummer sphere, 39
point mass, 33, 37
Poisson's equation, 33, 34, 36, 40, 56,
    129, 161
potential/density pairs
    fundamental cases, 36–8
    for Hernquist profile, 39
    for homogeneous sphere, 37
    for isothermal sphere, 37–8
    in Jaffe model, 38
    for King profile, 40
    for Navarro-Frenk-White (NFW)
        profile, 38–9
    for Plummer sphere, 39
    for point mass, 37
    for Yukawa potential, 40
Press-Schechter methodology, 140, 141, 170

redshift, 12–13, 122
relaxation time
    collisions and, 51
    definition of, 45

effect of integration limits, 47
hyperbolic trajectory, *49*
linear trajectory, *46*
of typical gravitating systems, **48**
remnant mass, 110
returned fraction, 110
Rosenbluth potential, 68

Salpeter IMF, 109
scale factor, 122, 123
Schechter function, 8–9
Schmidt's law, 17, 111
Sérsic index, 97
specific angular momentum, 93
spectral resolution, 24, 25
spiral arms
 Lindblad resonances, 103–5, *104*, 105, *105*
 morphology of, 100–1, *102*
 numerical simulations of, 106
 pitch angle of, *102*, 102–3
 rotating patterns of, *105*
 separation between, 103–4
 winding-up paradox of, 101–3
spiral galaxies, *7*
star formation rate (SFR), 9–10, 16, 17,
 111–12
star motion
 angular frequency of, 103
 asymmetric drift, 82, 88–9
 in cylindrical coordinates, 81
 epicyclic, 78, *79*, 80, *81*, 81–3, 103
 isolating *vs.* nonisolating integrals of,
 53–4, *54*
 random, 106
 rotational, 74, *75*, 75–7
 tangential, 80
 types of, 74
 vertical, 83–5, *84*, 86–7
stellar clusters
 binary, 154–5
 categories of, 22, 149
 core collapses of, 152–3
 definition of, 22
 ejection of unbound stars from, *160*
 escape velocity issue, 159–60
 evaporation of, 159–62
 evolution of density in, 153, 158–9
 expansion of, 151–2
 internal energy in, 154
 King models, 161–2
 mass loss of, 151–2
 mass segregation effect in, 153–4
 overview of, 149–50
 post core-collapse evolution of, 154
 steady tidal forces in, 155–7, *156*

tidal shocks, effect of, 157–9, *158*
stellar feedback, 18–19
stellar systems
 as collisionless ensemble of particles, 51–3
 collisions in, 47, 65–6
 distribution function, 44–5, 55–9
 dynamical friction effect in, 66–9
 isolating integrals of motion in, 53–4, *54*
 local and distant encounters in, 48–51
 motion of stars in, 46, 51–3
 nonisolating integrals of motion of, 53, *54*
 relaxation process of, 45–6, **48**, 50–1
 statistical treatment of, 43
 surface mass density, 57, 58, 84, 162
 two stars interaction in, *49*, 50
 See also galaxy clusters; stellar clusters
stellar yields, 108, 111
submillimetre galaxies (SMGs), 2
Sun
 distance from the galactic centre, 71
 metallicity of, 108
 motion around the galactic centre, 74, 75
 Oort's constants in neighbourhood of,
 76, 77
 orbital speed, 28
supermassive black hole (SMBH), 2, 11, 19
supernovae, 118, 119
surface brightness of galaxies, 5, 6, 92, 95–6,
 98, 129

third isolating integral, 86–7
tidal shock, *158*
Tolman dimming, 129
Tully-Fisher relation (TFR), 12, *95*, 95–6
2D projection, 41

Ultra-Luminous Infrared Galaxies (ULIRGs), 2
Universe
 age of, 127
 density of, 14
 Einstein-de Sitter model of, 121, 125–6,
 129, 130, 133, 134, 137, 138
 expansion of, 14, 121, 122, 123–4, 125
 flat, 122, 125
 Hubble's law of, 42, 123, 128, 129
 matter/energy content of, 14, 121

Vega (α Lyr), 22, 23
velocity dispersion, 45
virial theorem, 64–5, 96, 97, 151, 158

warm dark matter (WDM), 145
Wolf-Rayet stars, 151

Yukawa potential, 40

CPSIA information can be obtained
at www.ICGtesting.com
Printed in the USA
LVHW080404040320
648901LV00006B/27